THE HERITAGE SERIES

Samuel Burdick, Jr., c. 1880s–1890s.

From Madison to Mobile: The Diary of Corporal Samuel Burdick, Jr.

Company D, 23rd Wisconsin Infantry Regiment

Edited with notes by

DANIEL SCHARFENBERG

CORNERSTONE PRESS
UNIVERSITY OF WISCONSIN-STEVENS POINT

Cornerstone Press, Stevens Point, Wisconsin 54481
Copyright © 2026 Daniel Scharfenberg
www.uwsp.edu/cornerstone

Printed in the United States of America.

Library of Congress Control Number: 2026940952
ISBN: 978-1-968148-71-3

Cornerstone Press titles are produced in courses and internships offered by the Department of English at the University of Wisconsin–Stevens Point.

DIRECTOR & PUBLISHER
Dr. Ross K. Tangedal

EXECUTIVE EDITORS
Jeff Snowbarger, Freesia McKee

EDITORIAL DIRECTOR
Brett Hill

SENIOR EDITORS
Paige Biever, Lhea Owens

PRESS STAFF
Karlie Harpold, Sam Bjork, Sophie McPherson, Elizabeth Kyser, Andrew Bryant, Hannah Rouer

To my Grandparents Thomas Ames Scharfenberg and Judith Ellen Burdick, along with the rest of the Scharfenberg-Burdick family.

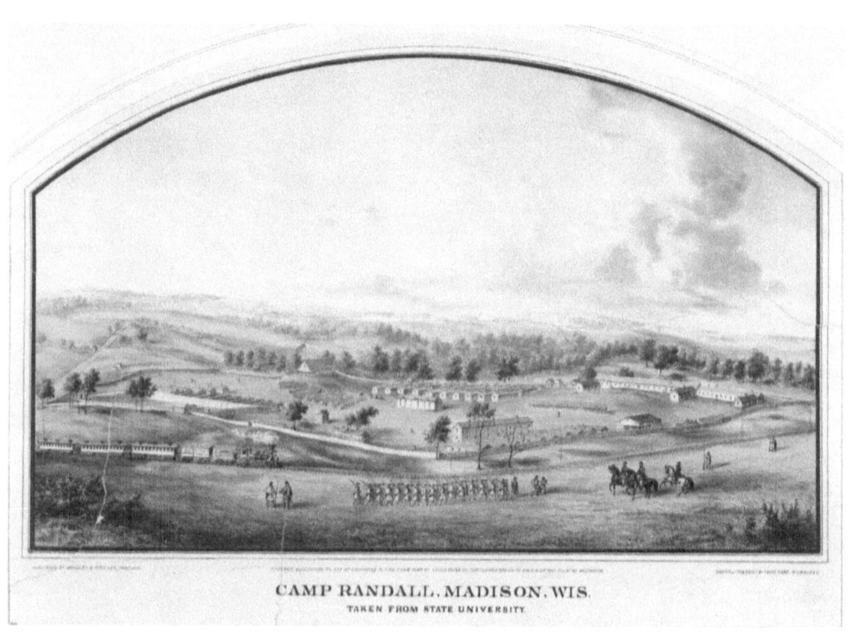

Camp Randall 1861-1865.

Contents

Introduction

Daniel Scharfenberg

Ever since I was young, I knew that one of my ancestors, my 3rd Great Grandfather Albert C. Burdick, was a veteran of the American Civil War. Beyond that however, I knew precious little about his involvement in the war and his experiences, that would not come until much later in my life while researching him. One day in the spring of 2020 while studying the paternal side of my family I noticed that one of my 3rd Great Grandfather's brothers was also a veteran of the American Civil War: this man was Samuel Burdick Jr., Albert's elder brother.

While browsing Samuel's ancestry to look for his Civil War service, I noticed that Samuel had a diary which was held at the Wisconsin Veterans Museum in the state capital of Madison. I quickly emailed a request for a digitized copy of Samuel's Civil War diary to their outreach archivist, Russell Horton, and patiently waited for a response in the mail. When a copy of the diary finally arrived in the mail, I was surprised at how immaculate Samuel's penmanship was and how easy it was to read. Minus a few grammatical errors and phonetic spelling, I quickly thought up the idea of having the diary published, as few resources exist on Samuel's unit, the 23rd Wisconsin Infantry Regiment. This book is not only a unit history of the 23rd Wisconsin but first and foremost it is a personal narrative written directly by a soldier in the Union Army from Wisconsin.

Samuel Burdick Jr. (1834-1914) wrote this diary from his time in the Union Army from 1862 to 1865. As previously mentioned, Samuel was the elder brother of my 3rd Great-Grandfather, Albert C. Burdick (1840-1919), both of Hopkinton, Rhode Island, making him my 3rd Great-Granduncle. The Burdick family had been in Rhode Island since Robert Burdick (1630-1690) first immigrated to New England and helped establish the settlement of Westerly, Rhode Island in 1680. Robert Burdick was originally from Devon,

England and came to British America as a freeman and fathered a total of 12 children with his wife Ruth Hubbard (1640-1691). Each of Robert's descendants went on to either travel westward or otherwise settle in various regions of upper and lower New England. The broader Burdick family is scattered across the whole of the United States, however, the majority of the Burdick family is still concentrated primarily in the Eastern United States and the Upper Midwest.

One of the 3rd Great-Grandsons of Robert Burdick was Samuel Burdick Sr. (1809-1866), also of Westerly. Samuel was the son of James Chesebrough Burdick (1777-1844) and Sally Ann Lewis (1799-1863) of Stonington Point, Connecticut. Samuel would marry Mercy Griffin Crandall (1810-1866) of Rockville, Rhode Island. Mercy's family, the Crandalls, were another prominent Rhode Island family going back to Elder John Crandall, an English Baptist Minister and founding settler of Westerly, of whom Mercy was his 4th Great-Granddaughter. Samuel and Mercy together had a total of 8 children, two girls and six boys, those being: Penelope Ann Burdick (1830-1850), John Monroe Burdick (1832-1906), Samuel Burdick Jr. (1834-1914), William Henry Burdick (1835-1910), Lafayette Washington Burdick (1838-1907), Albert Clarke Burdick (1840-1919), Marcus Morton Burdick (1843-1917), and Juliet Francis Burdick (1845-1905). Samuel was the second eldest of the six boys and the third eldest child overall.

As a child Samuel worked at the Upper Rockville Mill in the nearby settlement of Rockville, Rhode Island in the textile trade in the 1850's as a "card topper". Samuel's entire family later moved across the Pawcatuck River to Old Mystic, Connecticut just west of Hopkinton. Samuel and his brothers would all become ship carpenter apprentices at the Greenman Brothers shipyard in Greenmanville where Samuel worked until the late 1850's. In his early twenties Samuel moved west to the state of Wisconsin, likely in 1859 according to his postwar pension records, and settled in the village of Albion just southeast of the state capital of Madison near Lake Koshkonong. While in Albion Samuel lodged with Giles Franklin Lawton (1817-1868) of Verona, New York and the Lawton family. While in Albion Samuel worked as a carpenter in the village center and later owned a small plot of land just northwest of the village near his cousin, Henry C. Babcock (1805-1884), who was a 4th cousin of Samuel Burdick's family. Many of the landowners in Albion were extended cousins of one another, some of whom married distant family. Besides his involvement in the carpentry trade, Samuel was a dedicated Christian of the local Seventh Day Baptist Church just before the American Civil War commenced in 1861.

The village of Albion, Wisconsin was first settled by Freeborn Sweet (1810-1881), a New England pioneer from Oneida County, New York in August of 1841.[1] The settlement was named after the city of Albion, New York and was settled by a multitude of New Englanders, primarily from New York, Rhode Island, Massachusetts, and Connecticut, with a minority of the populace being new European immigrants to Wisconsin.[2] The city of Albion is surrounded by rich countryside and was primarily a tobacco farming community with a trading hub centered in the village. Albion, although a small rural settlement, was a deeply religious, educated, and patriotic community by the 1860's. Many of Albion's inhabitants were of either the Seventh Day Baptist or Primitive Methodist sects of Protestant Christianity which many of the New Englanders brought as a spiritual tradition with them from their homes back in the east.

According to prominent local educator Rasmus Bjørn Anderson, Albion was described as being the "ideal village, removed from all the temptations that surround young people in the cities".[3] Due to their strict religious policies, as well as their education, much of the populace of Albion supported temperance from alcohol and never served it to the populace.[4] Despite Albion being a rural farming community, much of the village was well educated. In 1861 Albion had a total of five schoolhouses spread throughout the village and surrounding farm settlements.[5] Albion was also the home of the Albion Academy, which at the time was one of the most prestigious educational institutions in southern Wisconsin.

The Albion Academy was founded in 1854 and taught both four and six year-long programs for students interested in furthering their education from across the state. Courses offered included Greek, Latin, vocal and instrumental music, elocution, mathematics, metaphysics, science, bookkeeping, and penmanship.[6] Many of the teachers for the surrounding schoolhouses were also taught at the academy. Business was contained primarily to the city center of Albion, which included several blacksmiths, carpenters, a saddlery, a dry

---

[1] William J. Park, *Madison, Dane County and Surrounding Towns; Being a History and Guide to Places of Scenic Beauty and Historical Note*, (Madison: W. J. Park & Co., 1877), 282.

[2] Park, *Madison, Dane County and Surrounding Towns*, 290.

[3] Rasmus B. Anderson, *Life Story of Rasmus B. Anderson*, (Rasmus B. Anderson: Madison, 1915), 76.

[4] Anderson, *Life Story*, 77.

[5] Ligowsky, A. Map of Dane County, Wisconsin. Madison, Wis.: Menges and Ligowsky, 1861.

[6] Park, *Madison, Dane County and Surrounding Towns*, 284.

goods general store, one hotel, as well as coopers, wheelers, and other such trades.[7] Samuel's listed profession in 1860 was still listed as a ship carpenter, the same trade which he had learned with his brothers while living in eastern Connecticut. According to the 1860 Federal Census for the city of Albion Samuel lived with the Lawton family, more than likely as a lodger or boarder in their house, as he had not yet established his own home in Wisconsin. Alice "Maria" Lawton (1831-1899), the wife of Giles Fraklin Lawton, was originally from Rockville, Rhode Island, a suburb of Samuel's hometown of Hopkinton. As their birth years are very close to one another, they were more than likely acquainted before Samuel's move to Albion, likely being classmates or neighbors.

Samuel and several of his brothers would go on to fight for the Union Army in the American Civil War, each in a different regiment, all of which would survive their respected campaigns, however, each of the Burdick brothers fought in completely different theaters of the war and had entirely different experiences and outlooks because of it. For example, Samuel's younger brother Albert fought as a Private, Corporal, and later Sergeant of Company G of the 5th Connecticut Infantry Regiment before being commissioned as the 1st Lieutenant of Company H. Albert eventually became the Regimental Adjutant as a staff officer for his regiment. Albert retired from the army at the closing of the war with the brevet rank of Captain. Albert was involved in many of the famous engagements of the Eastern Theater as part of the Army of the Potomac including Front Royal, Second Winchester, Second Bull Run, Cedar Mountain, Chancellorsville, Gettysburg, Resaca, Kennesaw Mountain, Sherman's March to the Sea, and ended the war fighting in the Carolinas at the Battle of Bentonville. Meanwhile, Samuel fought as a Private and Corporal in Company D of the 23rd Wisconsin Infantry Regiment in several key engagements of the Western Theater, including the defense of Cincinnati, the Siege of Vicksburg, the Battle of Arkansas Post, anti-guerilla warfare in Louisiana, Arkansas, and Mississippi, the Siege of Fort Morgan, and the Siege of Spanish Fort in Alabama. Both brothers were forever changed by their experiences of the war and this diary is a direct testament to that.

It is arguable that much of the Western Theater of the American Civil War is often overshadowed by the Eastern Theater for its memorable romanticized engagements, as well as well-known Generals of the American Civil War, both due in part to popular histories and media. Meanwhile, the Western Theater and its many engagements seem to be discussed with much less vigor or importance by scholars and the media alike. The Western Theater

---

[7] Park, *Madison, Dane County and Surrounding Towns*, 289.

of the war however was one of the most critical components to severing the Southern Confederacy in half via the Mississippi River to achieve Union General Winfield Scott's "Anaconda Plan" in its fullness. It is my hope that this diary highlights the importance of the theater and the hardworking worn-out lifestyle that was the life of the common Union soldier while on campaign during the American Civil War.

Camp Randall from the Northwest by Corporal John Gaddis of Company E, 12th Wisconsin Infantry Regiment. *Courtesy of the Wisconsin Veterans Museum.*

1862

INTRODUCTION

Just three days after the American Civil War broke out on April 12, 1861 after the bombardment of Fort Sumter, President Abraham Lincoln issued a call for 75,000 militiamen from all states to serve for 3 months service to quell the rebellion. These volunteers were nicknamed the "hundred days men" for their roughly 100 days of service.[1] By the summer of 1862 the American Civil War had already been raging for one full year across the United States. Although only one year into the war, the war itself began to encroach on the border of the American Midwest, primarily in states such as Ohio and Indiana. Deeming the initial 75,000 nationwide militia as inadequate to quell the rebellion, President Lincoln called for a further 42,034 more volunteers, this time to serve for a minimum of 3 years unless otherwise discharged.[3] Later, in July of 1862 the United States Congress approved 500,000 more volunteers to be enrolled into the ranks of the growing Union Army.[2] 500,000 more men would be appropriated by Lincoln just 2 weeks later that had to serve a minimum of 3 years.[3] Samuel Burdick falls into the third group of Union volunteers who volunteered for a minimum of 3 years' service with Company D of the 23rd Wisconsin Infantry Regiment. It should be noted that the majority of soldiers fighting in the war were volunteers as opposed

---

[1] William C. Davis, *Rebel & Yankees: The Battlefields of the Civil War*, (London: Salamander Books Ltd, 1991), 14; Ronald Paul Larson, *Wisconsin and the Civil War*, (Charleston: The History Press, 2017), 23; Lorenzo Thomas, *General Orders No. 15*. War Department, Washington, D. C, 1861.

[2] Leonard L. Lerwell, *The Personnel Replacement System in the United States Army.* (Department of the Army, 1954), 42.

[3] Ibid, 42.

to draftees, many of these volunteers fought for a variety of reasons, one such reason was the bounty system.

Military bounties were an incentive system which paid a lump sum of money, either from the county or state to citizens for American service as far back as the American Revolution and was prominent throughout much of the American Civil War on both sides.[4] Bounties varied wildly and were entirely dependent on the county and state of enlistment, some being as low as $10 and some as high as $300.[5] Private Wilbur Fisk mentions in his dairy the unfairness of bounties stating "two fellows might enlist from the same company, one would get a bounty that in all would amount to $702 at least, and perhaps more, while another who enlists to do the same duty in the same place, and whose name counts just the same to the credit of Vermont, get only $402. It is a strange chance game and very unjust".[6] In contrast to volunteers, men who were drafted were given no bounty at all. Bounties quickly became a popular way to get quick money, the men who practiced such thievery quickly became known as "bounty-jumpers". Bounty-jumpers were men who enlisted to pocket large bounties and would immediately desert the county or state, and then reenlist in a different regiment to claim another bounty.[7] Such practices went unchecked for much of the war well into 1865. Although the majority of Union troops were volunteers, there was a general fear since the war's beginning of a military draft.

The first draft of the war was not until 1862 with the passing of the Militia Act of 1862, and later the Enrollment Act of 1863, the Enrollment Act being America's first national conscription act.[8] According to *The Oxford Companion to American Military History*, "Four federal drafts produced only 46,000 conscripts and 118,000 substitutes (2 and 6%, respectively, of the 2.1 million Union troops). Most soldiers were U.S. Volunteers. However, the draft was credited, along with $600 million in enlistment bounties, with prodding volunteers, encouraging reenlistments, and demonstrating political will".[9] Aside from the draft was military service via substitution. Substitution was typically reserved for more affluent middle and upper class men who could

---

[4] Eugene Converse Murdock, *Patriotism Limited, 1862-1865; the Civil War Draft the Bounty System*, (Kent: Ohio State University Press, 1967), 16-18.

[5] John B. Billings, *Hardtack and Coffee: The Unwritten Story of Army Life*, (Boston: George M. Smith & Co., 1887) 215.

[6] Wilbur Fisk, *Hard Marching Every Day: The Civil War Letters of Private Wilbur Fisk, 1861-1865*, (Lawrence: University of Kansas Press, 1992), 186.

[7] Murdock, *Patriotism Limited*, 81.

[8] Ibid., 6-7, 63.

[9] Chambers, *The Oxford Companion to American Military History*, 181.

afford to buy their way out of a local draft lottery, typically this fee was $300 per lottery drawing.[10] Wealthier draftees could also offer to pay a friend or colleague a fee to serve in their place.[11] The Enrollment Act was extremely unpopular in the North, as the majority of men in the Union Army were volunteers, only a small number of men for the Union were draftees, and even less were substitutes.[12] The unpopularity of the Enrollment Act and its subsequent draft reached its peak in July of 1863 with the New York City draft riots which were met with violence throughout the city, primarily among the lower and middle class who increasingly saw drafts as a "rich man's war, but a poor man's fight".[13]

At the outbreak of war, the state of Wisconsin was already receiving hundreds of thousands of eager volunteers ready and willing to enlist in the Union Army. Adjutant General Lorenzo Thomas's General Order No. 15 called for each volunteer regiment to furnish at the absolute minimum: 1 Captain, 1 First Lieutenant, 1 Second Lieutenant, 1 First Sergeant, 4 Sergeants, 8 Corporals, 2 Musicians, 1 Wagoner, and 64 Privates.[2] According to Robert Sobel's *Biographical Directory of the Governors of the United States* "Lincoln's first call to arms asked Wisconsin for one regiment of ten companies, approximately 800 men".[14] Sixth Governor of Wisconsin, Alexander Williams Randall, who was a staunch supporter of Lincoln and a fervent abolitionist, followed through on his promise and raised not just one, but five total regiments of Wisconsin volunteers for the initial Wisconsin militia in 1861.[15] The 1st Wisconsin Infantry Regiment and the 2nd Wisconsin Infantry Regiment were the only two Wisconsin regiments to see action in 1861.[16] As Governor Randall saw the possibility of a protracted war he was able to procure millions of dollars to form more regiments and establish camps and forts across the state of Wisconsin to recruit and train volunteers.[17] By

---

[10] Fisk, *Hard Marching Every Day*, 149.

[11] Billings, *Hardtack and Coffee*, 215.

[12] Ibid.

[13] Murdock, *Patriotism Limited*, 230

[14] Sobel, *Biographical Directory of the Governors of the United States*, 1724.

[15] Robert Sobel and John W. Raimo, *Biographical Directory of the Governors of the United States, 1789–1978, Volume IV*, (Westport: Meckler Books, 1978), 1724.

[16] Richard N. Current, *The History of Wisconsin, Vol II: The Civil War Era*, (Madison: Wisconsin Historical Society, 1976), 296; Larson, *Wisconsin and the Civil War*, 24, 31-33.

[17] Larson, *Wisconsin and the Civil War*, 1724.

the end of 1861 Wisconsin had raised 16 full regiments of infantry.[18] A total of 53 volunteer infantry regiments, 4 volunteer cavalry regiments, 13 light artillery batteries, 1 heavy artillery battery, and 1 company of Berdan Sharpshooters were furnished by the state of Wisconsin from 1861-1865.[19] A total of roughly 91,379 Wisconsinites would serve in the American Civil War, of which 10,863 would be killed in action or died of disease.[20]

The 23rd Wisconsin Volunteer Infantry Regiment, otherwise known as the 23rd Wisconsin, had its origins in the Wisconsin state militia and was initially made up of 10 combined companies of various county-level militias in and around the state capital of Madison. The men of the 23rd Wisconsin came primarily from central and southern Wisconsin from the counties of Columbia, Dane, Iowa, Lafayette, Marquette, Sauk, and Waushara.[21] According to Edwin Bryant Quiner's *Military History of Wisconsin* the 23rd Wisconsin "was composed of four companies from Dane County, three from Columbia County, two from Sauk County and one from Lafayette County".[22] According to Charles E. Estabrook's *Annual Reports of the Adjutant General of the State of Wisconsin* at the time of the regiments muster in late August of 1862 the 23rd Wisconsin was "the best drilled of any from the state". [23] The 23rd Wisconsin was mustered into service at Camp Randall in Madison on August 30, 1862. Camp Randall was a military fort built in Madison in 1861 and named after the aforementioned sixth Governor of Wisconsin, Alexander Williams Randall.

Unit	Earliest Moniker	Recruitment Location	Earliest Captain
Company A	Madison Zouaves	Madison and Dane	William F. Vilas
Company B	Lafayette Guards	Lafayette	Charles M. Waring
Company C	Portage Guards	Columbia	Edgar P. Hill
Company D	Guppy Guards	Madison and Dane	Joseph E. Green
Company E	Lewis Guards	Dane	James M. Bull

---

[18] Ibid.

[19] J. D. Beck, *The Blue Book of the State of Wisconsin*, (Madison: Democrat Printing Company, 1907), 822-823.

[20] Ibid, 822-823; Larson, *Wisconsin and the Civil War*, 183.

[21] Charles E. Estabrook, *Annual Reports of the Adjutant General of the State of Wisconsin for Years 1860-1865*, (Madison: Democrat Printing Co., 1860), 183.

[22] Edwin Bryant Quiner, *The Military History of Wisconsin: a Record of the Civil and Military Patriotism of the State, in the War for the Union, with a History of the Campaigns in Which Wisconsin Soldiers Have Been Conspicuous, Regimental Histories, Sketches of Distinguished Officers, the Roll of the Illustrious Dead, Movements of the Legislature and State Officers, etc.*, (Chicago: Clarke & Co., 1868), 135.

[23] Estabrook, *Annual Reports of the Adjutant General*, 138.

Company F	Baraboo Rifles	Sauk	Charles H. Williams
Company G	Columbus Guards	Columbia	James F. Hazelton
Company H	Lodi Badgers	Lodi and West Point	Edmund H. Irwin
Company I	Capitol Guards	Madison and Dane	Anson R. Jones
Company K	Sauk Rifle Rangers	Sauk	Nathaniel S. Frost

Looking at the table above, Samuel's unit was nicknamed the "Guppy Guards". The Guppy Guards was the company moniker for Company D of the 23rd Wisconsin under the command of Captain Joseph E. Green. Captain Green was later promoted to the rank of Major and was the acting commander of the 23rd Wisconsin from January 1864 to June 1864, and again from January 1865 to June 1865. Company commanders at the time were often elected by their men rather than appointed, often being prominent men from their communities. Meanwhile, company-level monikers or nicknames were common at the time and gave a sense of comradery and individuality to different units. Monikers usually represented where the men of the unit were recruited from or, in some cases, were named after the company commander or regimental commander. Company D was primarily composed of men from Dane County. The list of cities and settlements listed within Company D's roster includes the cities of Madison, Pleasant Springs, Blue Mounds, Cottage Grove, Albion, Fitchburg, Berry, and Dunn. Other settlements in and outside of Dane County where a minority of soldiers from Company D were recruited from include the cities of Springdale, Sumner, Mazomanie, Argyle, Fulton, Arena, Cross Plains, Dane, Lafayette, Walworth, and Janesville. [24] In the case of Company D's nickname, the unit's namesake was for its regimental commander, Colonel Joshua James Guppey of Portage, Wisconsin, also sometimes spelled phonetically as "Guppy".

Joshua James Guppey (1820-1893) was a notable Wisconsin pioneer, lawyer, and military officer within Wisconsin state history. Guppey was born in Dover, New Hampshire in 1820. He was well educated, graduating from Dartmouth College in 1843 and was later admitted to the New Hampshire bar in 1846.[25] Guppey later moved to the frontier state of Wisconsin and set up a law practice in the city of Portage where he became the county judge of Columbus County from 1849 to 1857, he was also the Colonel of the Columbus County Militia shortly before the outbreak of war.[9] Guppey was commissioned as a Lieutenant Colonel in the 10th Wisconsin Volunteer

---

[24] Jeremiah M. Rusk and Chandler P. Chapman, *Roster of Wisconsin Volunteers, War of the Rebellion, 1861-1865. Volume II.*, (Madison: Democrat Printing Company, 1886), 241.

[25] Dorothy G. McCarthy, "The Contributions of Joshua Guppey," *Portage Daily Register*, January 22, 1972, sec. 2.

Infantry Regiment on September 13, 1861.[26] Guppey served with the 10th Wisconsin in 1861 for a year, skirmishing and ripping up railroads in Kentucky before being promoted to full Colonel and given command of the newly created 23rd Wisconsin Infantry Regiment on July 15, 1862.[10] Guppey served as the 23rd Wisconsin's regimental commanding officer for the majority of the war, falling ill on several occasions and leaving his second in command Lieutenant Colonel William Freeman Vilas as the temporary regimental commander until Vilas' resignation from the Army in 1863.[27]

When mustered out of service Guppey received the brevetted rank of Brigadier General, as ranks were commonly given in lieu of a military decoration.[28] After the war Guppey continued his career as the county judge of Columbus County for 15 years from 1865 to 1881, Guppey was also the superintendent of Portage schools from 1866 to 1871.[11] Guppey continued to serve in the Wisconsin State Militia, later the Wisconsin National Guard, until he retired in 1893.[11] Guppey died on December 8, 1893 from a combination of influenza and pneumonia in Portage, Wisconsin.[11] Joshua James Guppey is buried at Pine Hill Cemetery in his hometown of Dover.[29] Another major officer who led the regiment relevant to Wisconsin state history was William Freeman Vilas.

William Freeman Vilas (1840-1909) was a prominent lawyer and politician within Wisconsin. Born in Chelsea, Vermont Vilas' family moved to Wisconsin in 1851 where his father, Levi Baker Vilas, became a state assemblyman and the mayor of Madison. Vilas had served in the Wisconsin-based Governor's Guard militia and was the Captain of the Madison Zouaves militia, which later formed the bulk of the ranks of Company A of the 23rd Wisconsin. Company A was nicknamed after the Zouaves of the French colonial North African light infantry.[30] Vilas served as a Captain from August 1862 to February 1863, later being promoted to the rank of Major in February of 1863 and Lieutenant Colonel just one month later. Vilas temporarily took command of the regiment from June 5 to August

---

[26] Jeremiah M. Rusk and Chandler P. Chapman, *Roster of Wisconsin Volunteers, War of the Rebellion, 1861-1865. Volume I.*, (Madison: Democrat Printing Company, 1886), 644.

[27] Jeremiah M. Rusk and Chandler P. Chapman, *Roster of Wisconsin Volunteers, War of the Rebellion, 1861-1865. Volume II.*, (Madison: Democrat Printing Company, 1886), 231; Quiner, *The Military History of Wisconsin*, 713.

[28] Quiner, *The Military History of Wisconsin*, 719.

[29] "Gen. Joshua Guppy Dead." *Watertown Republican*, December 13, 1893. https://chroniclingamerica.loc.gov/lccn/sn85033295/1893-12-13/ed-1/seq-6/

[30] Merrill, *William Freeman Vilas: Doctrinaire Democrat*, 11-12.

25, 1863 while Colonel Guppey was incapacitated due to illness.[31] After his resignation in 1863 Vilas returned to his political and law career in Wisconsin, later becoming a senator for the state of Wisconsin, the 17th United States Secretary of the Interior, and the 33rd United States Postmaster General. Vilas was also an important postwar leader for multiple veteran organizations for the 23rd Wisconsin and major veteran reunions.[32]

A noteworthy point pertaining to 1862 relates directly to Samuel's future wife, Lucy Ann Saunders of Albion. In 1862 Lucy would lose two of her elder brothers, throughout the course of the conflict Lucy would eventually lose three of her brothers to the war.[33] This just goes to show the toll that the war had on just one family out of hundreds of thousands during the war. Lucy's eldest brother was Joseph Henry Saunders (1824-1862). Joseph served in Company H of the 1st Wisconsin Cavalry Regiment during the war. Joseph died at Cape Girardeau in Missouri of disease and is buried at the Evergreen Cemetery in Albion. Meanwhile, James Saunders (1837-1862) who Samuel knew before enlisting, joined Company B nicknamed the "Ripon Rifles" of the 4th Wisconsin Infantry Regiment. James died before Samuel's enlistment on April 6, 1862, and was later buried in the Chalmette National Cemetery in Louisiana. Lastly, Lucy's youngest brother George S. Saunders (1841-1864) served in Company I, the "Monroe County Volunteers" of the 4th Wisconsin Infantry Regiment. The 4th Wisconsin Infantry was later converted into a cavalry regiment and re-designated as the 4th Wisconsin Cavalry Regiment in 1863. George died in Baton Rouge, Louisiana, on December 14, 1864, and is also buried at the Chalmette National Cemetery.

A sidenote on the use of inst and ult throughout the diary text. Inst and ult are a mid-19th-century term often used to describe the current or previous month in Latin. The current month being "instant" or "inst", meaning present, while the preceding month is referred to as "ultimo" or "ult", meaning last. Samuel uses this term several times throughout the diary.[34]

---

[31] Quiner, *Military History of Wisconsin*, 711-713.

[32] Merrill, *William Freeman Vilas: Doctrinaire Democrat*, 12, 80-111, 133-138, 171.

[33] "Obituary – Burdick," *The Wisconsin Tobacco Reporter*, February 09, 1912, https://chroniclingamerica.loc.gov/lccn/sn86086586/1912-02-09/ed-1/seq-8/

[34] Merriam-Webster, *Webster's New Collegiate Dictionary*, (Springfield: G. & C. Merriam Company, 1977), 598, 1267.

ENTRIES

A journal kept by Samuel Burdick Jr. during three years of the war, commencing August 14th, 1862, Albion, Dane Co, Wisconsin.

Aug 14th, I went to a war meeting, at Albion Center, in the evening and after listening to patriotic speeches I, with several others, of my comrades, put our names down as volunteers for the United States Service and gave three cheers for the Union and went home.   Pleasant weather.

Aug 15th, I with the rest of the Albion boys started out for the City of Madison (this morning) for the purpose of being mustered into the U.S. Service. We went by train and arrived in the city at Eleven O'clock A.M. We took a stroll around the City and up into Camp Randall to see the sights, after which we went to the Capital. Toward evening we were sworn into the U.S. Service for 3 years unless sooner discharged by proper authorities, and was assigned to Captain Green's Co. (Co. D.) and was called the Guppy gaurds. We were then marched over to the Turner's Hall by Lieut. Stoltz where we staid over night. Pleasant weather.[35]

Aug 16th, We got a furlough this morning of the 10 days and started for home on the forenoon train, I remained at home until the 25th inst. doing up my work and setting up my business! I left my business in the care of Giles F. Lawton. The weather was pleasant generally.[36]

Aug 24th, E. S. Palmiter and myself went to Edgerton this evening and took the train for Madison our furloughs having expired. We arrived in Madison in due time and slept in the Depot building over night.[37]

Aug 25th, We went to Camp Randall this morning and were assigned quarters in the old Barracks, where we remained while in camp and were quite

---

[35] Lieutenant Francis "Frank" A. Stoltze (1832-1905) was one of the two 2nd Lieutenants of Company D. According to the company roster, Stoltze was a German from Madison. Stoltze resigned from active duty on February 7, 1863 for reasons of disability.

[36] Giles Franklin Lawton (1817-1868) was a fellow Seventh Day Baptist and very close friend of Samuel, taking care of Samuel's property while he was in military service, and was Samuel's landlord before the war. Samuel lived with the Lawton family for quite some time as a boarder in their home since 1859 when he first came to Albion. Giles and the Lawton family are all buried in Albion at the Evergreen Cemetery, as is much of the populace, including Samuel himself.

[37] Edgerton is a nearby town just south of Madison, Wisconsin and walking distance from the village of Albion. Many of Samuel's descendants later lived in Edgerton. Edwin S. Palmiter (1833-1862) was a friend and distant cousin of Samuel who served in Company D. Edwin died on December 27, 1862 of an infection

comfortable. Nothing of any importance transpired (except drilling) until the 30, when the other companys arrived in camp which were to compose the Regiment. Pleasant weather.

Aug 30th, The 10 Companies now being in camp we were mustered in as a Regiment. and numbered the 23rd Wis Volunteer Infantry and was paid 1 month's pay in advance and $25.00 on our Government bounty. Pleasant weather.[38]

Sept 1st, We remained in Camp Randall until the 15th inst. and drilled almost every day. I got a furlough of 2 days, during the time and went down to Albion and had a first rate good visit. Pleasant weather.

Sept 1st, We broke camp this morning and started for the seat of war. We left Madison on cars for Cincinnati Ohio, which city was threatened by an immediate attack by Braggs army. We arrived in Cincinnati on the morning of the 18th, after a pleasant trip of 2 days. We remained in the city nearly all day. The streets were thronged with citizens arrived with squirrel rifles and shot guns, n' had resolved to the defense of the city from all quarters of the state. Toward evening we crossed over the river under the command of Major Williams. We marched out back into the country about 4 miles, when it was ascertained that we were on the wrong road to the post that was assigned us and as it was dark we were ordered to climb up a steep bank close by, where we camped for the night. Pleasant weather and cool.[39]

Sept 19th, Our Regiment marched up the river this morning about 4 miles to the extreme left of our lines of defense and went to camp in a large peach orchard and named it "Camp Bates". Pleasant weather.[40]

Sept 20th, There was considerable excitement in camp today, caused by reports that the enemy were on the move to attack us in the night. Col. Guppy sent out a detachment of 70 men today on a scout (myself among the number) We soon ascertained that it was a false alarm, so we went home to camp all

---

[38] A bounty of $25 is roughly $740 today for enlisting as a volunteer. Such incentives were given to citizens soldiers to encourage them to volunteer, as opposed to being drafted.

[39] "Squirrel Hunters" was the nickname given to the hastily organized 15,000-man militia used to defend the city of Cincinnati in lieu of Union troops. The militia were nicknamed "squirrel hunters" due to their ragtag accoutrements and arms, most of which were antiquated flintlock, percussion muskets, and shotguns.

[40] Camp Bates was a Union military base near the Ohio border in northwest Kentucky. Camp Bates no longer exists and is poorly documented on maps, although a variety of soldiers diaries from various regiments do make mention of the camp. The installation was named after Ohio Brigadier General Joshua H. Bates of the Ohio State Militia who commanded a division during the time of the Defense of Cincinnati.

right. We remained in Camp Bates and drilled until October 4th without any further excitement as Bragg had commenced retreating out of state. Pleasant weather.

Oct 4th, We moved our camp today, out about one mile and in front of a large fort where we trained in camp and drilled until the 8th inst. when our division numbering about ten thousand more, started out on the march for Lexington, K.Y. for the purpose of driving the rebels out of that part of the state. Our division was commanded by A. J. Smith and our brigade by Green, Clay, Smith both of Kentucky. Pleasant weather and very dusty.[41]

Oct 9th, We broke camp this morning and marched 14 miles and went into camp for the night in a large meadow. This being our first days march with knap-sacks we were pretty tired and footsore. Pleasant weather and dusty.

Oct 10th We broke camp this morning at 6 o'clock and marched 10 miles and went into camp for the night at a place called Slaughter Valley where we did some foraging in the lewe of fresh meat and poultry. Pleasant.[42]

Oct 11th We broke camp this morning at 6 o'clock and marched 11 miles when we came to a town called Falmouth where we went into camp for the night. We remained in camp here until the 17th inst. Without much excitement. Except drilling. We had orders on the 15th to get ready to march with two days cooked rations. We got our teams all loaded and everything all ready to start when our Col. Came down from Division head quarters, with an order to remain in camp, which was very gratifying to us but which made it unpleasant for the most of the boys, after getting ready to leave they set fire to the straw and boards they had procured to sleep on, so they had to sleep on the ground until they could get more. Falmouth is a nice little town on the Covington Rail-road and no doubt was a brick business place before the war. We remained in camp here until the 17th. Pleasant and dusty.

Oct 17th, We broke camp early this morning and left Falmouth for Cynthiana where we arrived at 8 o'clock P.M. it being 23 miles and went into camp for the night in a large field wear a large fort, built by the home guards, before the place was captured by Bragg's army. Cynthiana is quite a large town and a very pleasant place.   Pleasant and dusty.

---

[41] Green Clay Smith and Andrew Jackson Smith were both Union Generals and commanders in the XIII Corps.

[42] Foraging is a period term for commandeering property, often food to be used by an Army from the civilian populace.

Oct 18th + 19th, We lay in camp and rested and washed up our clothes. Pleasant and cold.

Oct 20th, We broke camp this morning at Cynthiana and started for Paris. We marched 18 miles and went into camp for the night. Pleasant and cold.

Oct 21st, We broke camp this morning and marched to Paris, (it being 10 miles) and went into camp in a large grove. Paris is a very pretty place and quite a city. Pleasant and cold.

Oct 22nd, In camp at Paris. Last night there was some apprehension of an attack from the enemy and we slept on our arms during the night. Several deserters came to our camp from the rebels last evening and were kept under guard. Pleasant and cold.[43]

Oct 23rd + 24th, In camp at Paris we fixed up our huts today as it was very cold weather.

Oct 25th, This morning it commenced to rain and toward the evening it turned into a snow storm which made things look very uncomfortable for the night as we had no stoves in our tents. A very stormy night.

Oct 26th, This morning the snow was 6 inches deep and the weather very uncomfortable. Some Illinois Regiments came in from the North today and camped near us. Pleasant and cold.

Oct 27th, In camp at Paris. Gen. Green Clay Smith took leave of our brigade today and Gen. S. G. Burnbridge took his place in command and the brigade was newly organized Our brigade was composed of the 83rd + 96th Ohio Regiment, the 16th + 67th Indiana and the 23rd Wisc, and the 17th Ohio battery of 6 guns. Pleasant weather and cold.

Oct 28th, We broke camp early this morning and left Paris for Nicholasville. We marched 24 miles and went into camp in a large walnut grove 4 miles from Lexington. Pleasant and cold.

Oct 29th + 30th, We lay in camp to rest and wash up our clothes and crack old English walnuts. Pleasant weather and cold.

Oct 31st, We broke camp this morning and marched to Nicholasville, (it being 17 miles) and went into camp in a nice grove. We passed thru Lexington today. It was a large and very nice city. Cloudy weather.

---

[43] "Sleep on arms" was a command given to soldiers to keep all their accoutrements on their persons while sleeping and to be ready for immediate action. Arms could be held onto or stacked. This command was often given while waiting for an impending attack from the enemy and allowed any given unit ample time to organize into line of battle to repel an opposing force while still allowing soldiers to sleep.

Nov 1ˢᵗ, 2ⁿᵈ & 3ʳᵈ, We lay in camp and fixed up our quarters and washed up our clothes.

Nov 4ᵗʰ, Our Regiment voted for county officers this forenoon and went out on Division review in the afternoon. Pleasant weather and cold.

Nov 5ᵗʰ, 6ᵗʰ & 7ᵗʰ, We lay in camp and drilled a little and fixed up our tents. Everything all quite around camp. Pleasant and cold.

Nov 8ᵗʰ, Our Division was reviewed again today by our Division commander, A. J. Smith.

Nov 9ᵗʰ + 10ᵗʰ, We lay in camp and got ready for a march. Pleasant and cold.

Nov 11ᵗʰ, We broke camp this morning and started for Louisville, it being ascertained that Gen. Bragg's army had left this part of the state. We marched 17 miles and went into camp for the night, 2 miles from Versailles. Pleasant.

Nov 12ᵗʰ, We broke camp this morning and marched to Frankfort, the capital of the state. about 15 miles. We marched thru the city and went into camp for the night. Frankfort is a very dirty city. Cloudy.

Nov 13ᵗʰ, We broke camp this morning and marched 18 miles and camped for the night.

Nov 14ᵗʰ, We broke camp early this morning, marched 18 miles and went into camp for the night. Pleasant and cold.

Nov 15ᵗʰ, We broke camp early this morning and marched to Louisville, thru the city, and one mile beyond and went into camp, marching 16 miles. Pleasant weather and cold.

Nov 16ᵗʰ, 17ᵗʰ & 18ᵗʰ, We lay in camp at Louisville, washed up our clothes and drawed some new clothing. Pleasant and cold.

Nov 19ᵗʰ, We broke camp this afternoon in a rain storm, and after standing around in the rain and mud for an hour we marched down to Portland Landing on the Ohio River (it being about 1 mile) to take the boats for Memphis. 4 companies of our Regiment went on board the steam boat "Masonic Gem" and the other 6 went on board the "Sir Wm Wallace". We started down the river at about sundown, run about 10 miles and lay up for the night. It rained all day and night.[44]

Nov 20ᵗʰ, We started on down the river this morning but did not make much headway as the water was so low that we got grounded several times. We made about 60 miles and tied up for the night. Pleasant.

---

[44] Steamboats became one of the most utilized forms of river transportation during the war and were used prominently by both belligerents during the Vicksburg Campaign to transport troops and bombard enemy fortifications.

Nov 21ˢᵗ, We started out this morning and made 65 miles and tied up for the night.

Nov 22ⁿᵈ, We started on down the river this morning and got a ground. It took us about half of the day to get it off. We made 35 miles and lay up for the night. Pleasant.

Nov 23ʳᵈ, We started on down the river this morning and got a ground on Scuffle Shole Bar. We fooled around about half of the day. We passed Evansville to ~~Louis ville~~ day. The distance from Evansville to Louisville is 200 miles. We made about 40 miles today and tied up for the night. Cloudy and rainy.

Nov 24ᵗʰ, We started out early this morning and soon ran onto a snagg which caused our boat to leak considerable. We soon ran onto a sand bar and after fooling around for about 2 hrs we got off and started down the river We reached Smithfield at 11 A.M. Passed Paduca at 2 P.M. and arrived at Cairo at 8 O'clock in the evening. We made 75 miles today in all. Pleasant and cold.

Nov 25ᵗʰ, We left Cairo last evening at 10 O'clock and run all night. We passed Columbus at 12 O'clock in the night. We made about 146 miles last night and today and lay up for the night. Pleasant weather.

Nov 26ᵗʰ, We started out early this morning and arrived in Memphis at noon, (it being 110 miles). We got off the boat and marched out back of the city about 3 miles and went into camp for the night. Cloudy and rainy.

Nov 27ᵗʰ, 29ᵗʰ + 30ᵗʰ, We lay in camp and fixed up our quarters and washed up our clothes. Pleasant + cold.

Dec 1ˢᵗ, We broke camp today and marched back to the city and down the river about 3 miles and went into camp on the ground just vacated by the 32ⁿᵈ Wis. Regiment. They had winter quarters nearly completed. We took the logs and bricks that they used and fixed up our quarters comfortable. Pleasant and cold.

Dec 2ⁿᵈ, We remained in camp until the 17ᵗʰ inst. Without any excitement, except the regular routine of camp life, such as fixing up our quarters and drilling a little every day. Pleasant weather and cold.

Dec 17ᵗʰ, The army here was reviewed today by Gen. Sherman. He complimented our Regiment very highly. Pleasant.[45]

---

[45] William Tecumseh Sherman was a Union General in the Army of the Tennessee. At this point in time, Sherman commanded the right wing of the Union's XIII Corps.

Dec 18th, Capt. Frost of Co K. died today in the Hospital. Pleasant weather and cold. [46]

Dec 19th, A detail was made today from our Regt. of 100 men to go over to the river at the fort and load a Steamer with ammunition to go down the river. Pleasant weather and cold.

Dec 20th, Got marching orders today to get ready to go on board the boats to go down the river. Tore up our camp and sent our camp equipment and tents down to the landing and got ready to march at sunset. We got into line and started out, where the orders came that the boats had not arrived and that we were to stay in our old camp overnight. So we spread out our blankets and lay down for the night. Pleasant but very cold.

Dec 21st, Morning. we marched over to the forts and stacked arms where we remained until noon when we went on board the steamboat "John H. Dickey" and started off down the river with the rest of the fleet. Led off by two gunboats at about 3 oclock in the afternoon. We passed Helena at midnight and laid up about 12 miles below at a place called Friar's Point on the upper side of the river.

Dec 22nd, At day light about 40 boats loaded with troops were in sight at the point. Our men nearly burnt the town of Friar's Point in retaliation for injuries done our soldiers. We started out from Friar's Point at 2 O'clock P. M. and run all night. Pleasant weather and cold.

Dec 23rd, We passed Napolian today at 11 O'clock A.M. and run down to a place called Gaster's Landing (making 100 miles) where we lay up and waited for the ballance of the fleet to join us. Cloudy and rainy. [47]

---

[46] Captain Nathaniel S. Frost was commissioned as the Captain of Company K, nicknamed the "Sauk Rifle Rangers" of the 23rd Wisconsin on September 1, 1862. Frost was born in Norway, Maine on August 8, 1830, and according to the company roster was a resident of Prairie du Sac, Wisconsin before his commission. Corroborating with the company roster, Frost did indeed die at the young age of 33 on December 18, 1862, near the outskirts of Memphis, Tennessee from disease.

[47] Napoleon, Arkansas is a city at the converge of the Arkansas and Mississippi rivers and was occupied by Union troops by 1862. Much of its port-based economy was destroyed from the war, as well as the physical city from the flooding waters of the rivers. Likewise, Gaster's Landing was a waypoint and ship landing to offload both troops and supplies.

Dec 24th, Left Gaster's Landing early this morning and run down to a place called Milegen's Bend about 25 miles above Vicksburg, on the opposite side of the river, where we laid up for the night (making a run of 150 miles). Pleasant and cold.[48]

Dec 25th, Christmas day. Our Brigade Commander got orders this morning to land his Brigade and march out into the country 28 miles to a place called Dallas Station, on the Texas. + Vicksburg railroad and to destroy the bridge and track at that place, to cut off the rebel's supplies from Texas. We got ready and started on the march at 9 o'clock A.M. We arrived there at 9 o'clock in the evening without any opposition from the rebs and commenced to tear up the track. We worked until midnight and then lay down to sleep, wet, tired and hungry. It rained all day and night.[49]

Dec 26th, We got up at about 5 o'clock and got a good breakfast of chicken, sweet potatoes and honey which we found in abundance after which we commenced our work of destruction again. We set fire to the Station's buildings the railroad bridge and the ties that we had piled and after seeing them thoroughly destroyed we started back for the river at about 9 o'clock A.M. and arrived at the river at about 9 in the evening tired, wet and hungry, making a march of 56 miles in 36 hrs, beside tearing up over a mile of railroad track, etc. We burnt about 2000 bales of cotton and captured about 500 mules and horses and brot in with us. It rained all day and night.[50]

Dec 27th, Our boats started out this morning after those that had already done down the river. We run down to the mouth of the Yazou river and up it 8 miles where our other boats had landed. We got off the boats and marched out into the swamp to within 4 miles of Vicksburg, where we lay down to get some sleep, all being pretty wear tired out.   Pleasant and cold.

---

[48] *Miligen* here is spelled phonetically in reference to Milliken's Bend. Milliken's Bend was a landing on a bend in the Mississippi River between Louisiana and Mississippi just northwest of Vicksburg named after Major J. Milliken. Many Union troops frequented Milliken's Bend as an off-loading area and waypoint while maneuvering down the Mississippi towards Vicksburg throughout the campaign. Milliken's Bend was also the site of the Battle of Milliken's Bend in June of 1863.

[49] Critical to the Union's effort to take Vicksburg was to first destroy the Shreveport-Vicksburg railway which supplied much needed provisions such as beef, salt, sugar, and molasses coming from the western Confederate states of Texas and Louisiana. By destroying the railway and crippling Confederate supply lines it would sever Confederate supplies coming from the west.

[50] A popular form of destroying Confederate railways throughout the war was by heating up the iron rails over the wooden railroad ties, this would make the iron malleable from the heat and could be easily bent.

Dec 28th, We were aroused at about 5 o'clock in the morning by the roar of artillery in our front. We immediately got in readiness for a fight and soon after the volleys of musketry told us that a fight was going on at our left. At about 10 o'clock A.M. our Brigade marched out to the front about 1 mile and took position on the right of our line of battle in the timber and in the sight of the rebel batteries. Here we remained all day, the rebel shot and shell cutting thru the trees and whistling all around us but only 1 man was killed and a few wounded near us. Our position was on the low ground in the timber. At night we gathered up some leaves and brush to lay on. We lay in line of battle all night. The rebels occupied a range of hills in our front that were very strongly fortified. Pleasant weather and cold.

Dec 29th, We awoke bright and early this morning, refreshed from our nights rest. Soon after daylight the batteries opened fire quickly on to the sides. The heaviest firing, as the day before, being on our left, about a mile distant. As soon as we got breakfast we sent out two companys of skirmishers a long to a bayou that lay between us and the rebs. They were out about 2 hrs when they were ordered to join the Regiment again. We had moved back in the woods about 100 rods and were waiting for the skirmishers to join us when we got orders to march out and reinforce the left of our line that was having a severe fight. We had not gone far when we got orders to about face and march back to our old position as the rebs were getting the worst of it on our left. We threw out skirmishers again which acted as pickets thru the night. It commenced raining at about dark and rained smartly all night, thoroughly drenching us to the skin and making things very uncomfortable for the night. We did not get much sleep. There was no firing thru the night except an occasional heavy gun at regular intervals. The weather was cloudy thru the day and rained all night.

Dec 30th, It cleared up about daylight and at about 8 o'clock in the morning we could see a plenty of rebels walking along on top of the breastworks on the hill. At about 10 o'clock A.M. we were marched back about one mile to the rear and were allowed to built a fire to dry our clothes and cook some rations where we remained thru out the day undisturbed. There was but little firing on either side today and it seemed that there was general cesation of hostilities on both sides. Pleasant and cold.

Dec 31st, Everything is quiet this morning and apparently nothing is going on. The general impression is that the capture of Vicksburg will be attended with severe fighting and a heavy loss of life. Toward evening we fell in line and were mustered for two months pay. After dark we again fell in line and proceded quietly 1 mile to the front along the bayou that lay between us and the rebs, where we worked all night by reliefs throwing up rifle pits and breastworks. We worked until 4 o'clock in the morning when it began to get light and withdrew back into the 1 mile. The whistles of the locomotives in Vicksburg could be distinctly at all hours of the night and it was evident that they were busy about something Pleasant and cold.

Siege of Vicksburg by Thure de Thulstrup

1863

INTRODUCTION

1863 was a pivotal year in the American Civil War, both in the East and in the West. With General Ulysses S. Grant now in charge of the Army of the Tennessee, the war could now be pressed on northern and western Mississippi for full control over the Mississippi River effectively cutting the Confederacy in two. The city of New Orleans had already been captured by the Union Army and Navy in May of 1862. Linking Union troops via the Mississippi River to those in Louisiana became imperative to Union success in the west. To General Grant and others, the war could not be won without full uncontested control of the Mississippi River.

To achieve this end, the Union created the so-called "brown-water navy", a navy of ironclad and gunboat ships capable of traversing the Mississippi River and delivering destructive firepower to all forts that lay on the banks of the Mississippi. One such installation were the defenses and forts around the city of Vicksburg in Mississippi. These fortifications and the city itself would have to be sieged and captured in order to complete the strangulation of the broader "Anaconda Plan". The Anaconda Plan was originally proposed by Union General-in-Chief Winfield Scott in 1861. The plan consisted of blockading the Confederate states via ports, rivers, and waterways surrounding the south, thus denying the Confederacy any imports by slowly crushing the Confederate economy and trade. Achieving Scott's plan would come at the cost of nearly 11,000 Union casualties at Vicksburg alone. The siege of Vicksburg would see some of the heaviest bombardments and siege tactics of the entire war, the siege itself would ultimately end in a Union victory on July 4, 1863.

By January of 1863 Samuel and the 23rd Wisconsin had not seen any major fighting in the XIII Corps other than being in range of small skirmishes. In fact, more men in the unit had actually been killed by disease or in accidents than in battle. The skirmishes and battles preceding the Siege of Vicksburg, as well as the siege itself, would become one of the most important series of actions which the 23rd Wisconsin would take part in during the war in 1863 alone. While the 23rd Wisconsin was technically "engaged" during the Battle of Chickasaw Bayou, also called the Battle of Walnut Hills, from December 25 – 29, 1862, Company D was not in the formal battle as it was busy destroying the Vicksburg-Shreveport Railroad at Dallas Station. Samuel and Company D's "baptism by fire" would not occur until January 9 – 11, 1863 during the Battle of Fort Hindman, also called the Battle of Arkansas Post.

ENTRIES

Jan 1st, New Years Day. We remained quietly all day in the same position that we occupied yesterday until about 8 o'clock in the evening when we were ordered to fall in line and march off thru the woods and mud for about 2 hours without knowing our destination where our eyes were gladdened by the sight of our boats which were reached in safety at about 11 o'clock in the night. We lay down and slept on the river banks till morning.   Pleasant and cold.

Jan 2nd, When we got up this morning it was currently reported that the rebs had been largely reinforced from this point and that they were determined to hold Vicksburg at all hazards. Consequently we had to abandon withdraw our army and abandon the attack on Vicksburg as our forces were to small to carry the rebel works. We met quite a loss in falling back as the rebs followed us up. We got everything loaded onto the boats and started down the Yazou river at about 10 o'clock A.M., the gun boats covered our rear. So ended the battle of Chickasaw Bayou with a heavy loss on our side. We run down the Yazou into the Mississippi and over the opposite side where we lay up for the night. It rained all day and night with heavy thunder.

Jan 3rd, We lay at the bank all day and were allowed to go on shore. There was a great deal of speculation today as to what we were going to do next. It rained all day and night.

Jan 4th, Toward morning it cleared off cold and about the middle of the fore noon our boat run down along side of the Quartermaster's boat and took on some rations. Pleasant and cold.

Jan 5th, This morning the fleet started off up the river. We run up to a place called Ganes Landing, where we laid up for the night (60 miles above Vicksburg.) Pleasant weather.[1]

Jan 6th, The fleet started out up the river this morning we made about 35 miles and laid up for the day and night. Here we killed a supply of fresh beef, the first we have had in a long time. Pleasant and cold.

Jan 7th, We started on up the river this morning. at 5 o'clock. We had a gunboat in tow. We run all night. Pleasant.

Jan 8th, We arrived at the mouth of the White River this morning and tied up on the opposite side of the river where we lay all day and night (making 145 miles)   Pleasant weather.

Jan 9th, This morning, the fleet led off by the gunboats, started on up the White River until we came to the cut off, that makes into the Arkansas River. We run thru the cut off and up the Arkansas River to a point a short distance below Fort Hindman at Arkansas Post, where we made preparations for taking the Fort (making in all a 35 miles run). We lay on shore thru the night. Pleasant and cold.

Jan 10th, This morning the gunboats commenced the attack with the assistance of a part of the land forces, they drove the enemy out of their outer entrenchments and into their inner works. Our regiment having previously landed, was marched (at about noon) out thru the swamp and mud about 1 ½ miles towards the fort, where we halted till evening and then we moved up thru the woods into the rear of the fort where we lay on our arms until morning. Our march was continued up thru the woods and in the evening followed by a regular shower of shot and shell from the rebel fort but fortunately no one was hurt in our regiment. The cannonading was kept up all night by our gunboats and the fort. We lay on our arms thru the night and were very uncomfortable as our feet were wet and the night was very cold and we had only 1 blanket. Pleasant and cold.

Jan 11th, Sunday morning. At day light we marched about 1 mile up toward the fort where we halted again. The forenoon was spent mostly in getting our batteries and troops into position with out much firing on our side, while the rebels sent a regular shower of shot and shell in around us from the fort. The gunboats opened on the rebel fort at about 1 o'clock P.M. in

---

[1]   Gaines Landing was one of many stopping points along the rivers and canals for both Confederate and Union troops during the Vicksburg Campaign. Gaines Landing was located just south of Vandalia, Mississippi on the Mississippi River, the landing itself no longer exists.

good earnest and the fighting had begun on our right quite frisk. At 2 o'clock our regiment was ordered up to the front and we immediately started off on down quick which we kept up for about one half a mile which got us close vicinity to the fort. We advanced all the way under the very heavy fire from the fort. After the desired position had been gained, we halted for a few moments to get breath and reform our lines and while we were doing this, Gen. Burbridge rode along in front of our regiment and told us that one of his regiments had given back in confusion and that the fortunes of the day depended on us. Col. Guppy then gave orders to fix bayonets and advance on the fort. We immediately moved forward in line of battle and no sooner had we immerged from the woods and bush than we received a terrible cross fire from artillery and musketry from the rebel works, but fortunately for us, they fired too high the most of the shot and shell going over us. After several of our men had fallen, 2 companies were sent out on our left as skirmishers and they soon checked the rebel fire in that direction. We kept them down for a few minutes, when a shell came from the fort and struck in the center of our company, killing 3 men and wounding 7 more. We were then ordered to fall back into the bush a few rods and throw out a line of skirmishers who succeeded in keeping down the rebs so that our batteries could move up closer and get a better range when they gave the rebs grape and canisters in double doses. At about 3 o'clock P.M. Gen. Burbridge again ordered our regiment to form a good line and storm the fort. He said that it would be necessary to take the fort at this point of the bayonet and that he was sure that we could do it. He gave the order for us to advance on the fort when we started with a yell, determined to take it or die, but we had not got more than half way to it, across the open field when the rebels run up the white flag and surrendered the works to us. It was not many minutes before the rebel flag came down and the glorious old Stars and Stripes were floating in its place and the air resounded with Union cheers. Our troops were all marched into the fort where they stacked arms with the exception of 1 regiment, that was placed on guard over the prisoners. We were then allowed to go around and see the fort. We captured 6800 prisoners 22 pieces of artillery, 3 of them, 100 pounders and the rest of them, 6, 10 + 12 pounders. We also captured 8000 stand of small arms and all of their camp and garrison and a large lot of ammunition and provisions. Our gunboat and artillery did terrible execution in the fort. The ground was strewn with dead men and artillery horses. We dismounted almost every piece of their artillery and some of it was knocked to pieces. The rebels lost in killed and wounded about 700 and our loss was about the same. Our regiment lost 7 killed and 27 wounded. After looking the fort all over our regiment was marched out back of the fort about 1 miles into the woods and camped for the night. Gen. Burnbridge complimented our regiment very highly for the part it took in the fight. The weather was cloudy and cold.

Jan 12ᵗʰ, Morning. We lay in the woods until noon and after getting our dinner we were marched back to the fort and were stationed around it as guards We remained on guard until evening where we were relieved and marched down the river 2 miles, to the boats, through the mud and rain. We went on board the old John H. Dicky and lay on her over the night. Weather cloud and rainy.

Jan 13ᵗʰ, Went on slow this morning, lay around all day and washed our clothes and cooked some rations etc.   Pleasant and cold.

Jan 14ᵗʰ, On board the John H. Dickey. A detail of 100 men from our regiment was sent up the river today under the command of Capt. Green, of our company. on the tug boat Laren to get some corn that was known to be up there. About 20 miles on their way up they saw several parties of guerrillas. When they got up there, Capt. Green saw that he could not get the corn on board the boat on account of the guerillas, so he set fire to it. It had just got to burning nicely when a party of guerillas dashed out of the woods out there and made an effort to capture the boat, but our boys rushed onto the boat and cut the rope and as the boat had up stream they got off with out much loss on their own side, except 6 of our boys that had gone up to a house to get some chickens, so they were captured by the guerillas. There was 3 or 4 of our boys wounded on the boat. The boat made good time coming back down the river as the guerillas followed them along the bank to get a shot at them but our boys kept a good lookout for them. They got back about sunset all right. Pleasant weather thru the day, rained in the night and turned into a snow storm towards morning. and cold.

Jan 15ᵗʰ, Morning. The snow was from 4 to 5 inches deep this morning. In the after noon the most of the fleet started down the river, while a part of it waited at the fort for the troops that were detailed to destroy the fort as our General did not want to hold it. A part of the fleet run down the White river and another part down the Arkansas river. We run down the Mississippi and lay up on the opposite side at the mouth of the White River. Cloudy + cold.

Jan 16ᵗʰ, We lay opposite the mouth of the White River all day. In the evening a detail of 300 men was send out on a scout (50 of these from our regiment) under the command of Lieut. Col Jusen of our regiment, after some guerillas that were bound to be out about 5 miles from us. Our boys marched out thru the mud + water but before they got out there the guerillas has left so they had their trouble for nothing, and had to march back again thru the wet without accomplishing anything. Rainy and cold.[2]

Jan 17th, We run down to Napolian this morning at the mouth of the Arkansas river, where we found the ballance of the fleet and where we remained all day. Our Company buried two men here that had died on the boats. Pleasant weather.

Jan 18th, Sunday. We got some fresh meat today and cleaned up our clothes. Pleasant weather.

Jan 19th, 20th + 21st, We lay at Napolian all day. Pleasant.

The fleet started down the river this morning and arrived at Milligen's Bend at 5 o'clock P.M. and were making preparations to when the orders came for us to proceed down farther. We run down 12 miles and lay up for the night, opposite the mouth of the river (Yazoo) and in plain sight of Vicksburg, it being only 12 miles off.   Pleasant weather.

Jan 22nd, This morning the boats carrying our brigade run up the river about 2 miles and lay up and the men were allowed to go on shore, which they were all glad to do, after being on board the boats 4 weeks with exception of a few days. Our regiment were in rather hard condition at this time, being very dirty and most all sick.   Pleasant and cold.

Jan 23rd, We went on board the boats this morning and at about noon our boats again dropped down to the mouth of the Yazoo river and in the afternoon, a part of Gen. Grant's army arrived and we all lay on the boats over night. Pleasant + cold.

Jan 24th, We lay on the boats until about 3 o'clock in the afternoon when we got off and marched down the river about 3 miles and went into camp in plain sight of Vicksburg. Our wagon got stuck in the mud going down, and did not get there till morning, so we had to get along without our tents thru the night. The weather was pleasant and cold.

Jan 25th, We remained in camp until Feb 14th, and did some picket and guard and drilled a little after fixing up our camp and tents. The weather pleasant most of the time and cold.

---

[2] Edmund Jüssen was a German revolutionary, lawyer, and Wisconsin pioneer of both Columbus and Portage, Wisconsin. He was the Lieutenant Colonel of the 23rd Wisconsin Infantry Regiment until March of 1863 (see image in middle of book).

Feb 14th, Our brigade got orders this morning to go up the river to Greenville on a guerilla chase with 7 days rations. We arrived there at about 2 o'clock in the afternoon and landed and started out after the rebs. We chased them 14 miles but they being mounted and we being afoot, they got away from us, so we went into camp for the night on a large plantation where we got all the chicken, sweet potatoes and honey, that we wanted. It rained all day.

Feb 15th, We could not hear anything from the rebs this morning so we started back for the river where we arrived all night. We captured a rebel Col. And 7 men and also destroyed a large lot of cotton. It rained all day.

Feb 16th, Early this morning we started on up the river for a point about 30 mils below the mouth of the Arkansas river where the guerillas had been firing onto our boats. We landed at about 11 o'clock in the forenoon and started out after them We skirmished with them and our artillery, shelled them for some ways threw the woods, but they being all mounted soon got out of our reach. We chased them 15 miles when we came to a bayou where they that crossed and distroyed the bridge, so we shelled them across the bayou a spell and as it was nearly night, we went into camp for the night. We captured 15 rebs and one piece of artillery. We had 1 man killed and 3 wounded. The rebels had 7 killed and 12 or 15 wounded. I was on picket thru the night. We had no ration with us nor any blankets, and as it was a very cold night we were very uncomfortable before morning. Pleasant and very cold.

Feb 17th, Early this morning as we could not hear any thing of the rebs, we started back for the boat after destroying several large buildings and a large lot of cotton, with 17 prisoners and 1 piece of artillery. We arrived at the boats about noon and went on board and started on down the river. We got back down to Greenville in the evening and were informed by the gun boats that the rebs had come back, while we had been up the river, so the General landed the troops ready to chase them again in the morning. We lay on the shore during the night. Cloudy + cold.

Feb 18th, Early this morning we started out into the country to look after the rebs. They were about 1000 strong (all mounted) and 6 pieces of artillery. We had about 300 cavalry, 1500 infantry and six pieces of artillery. We soon got track of them and chased them about 15 miles where we came to a bayou where they had crossed and tore up the bridge after them. Our cavalry forded the bayou and still followed them but our artillery was detained about 2 hrs in getting across. The infantry forded the stream and and we started on after them. We followed them for 3 miles but they had got so much the start and it being night we got ordered to go into camp for the night on a large plantation where we foraged freely and had a jolly old supper and breakfast. Rained all day and very cold. There was no loss on our side.

Feb 19th, Early this morning, not hearing anything from the rebels, we started back to toward the boats. We foraged considerable and burnt a large lot of cotton and several large building. Captured a large lot of mules and horses, and beef cattle which we took back to camp. We got back to the boats at 4 o'clock in the afternoon. We lay on the bank over night. Rainy and cloudy.

Feb 20 + 21st, We lay at Greenville and had a good old time.   Pleasant.

Feb 22nd, We left Greenville this morning to go down the river to our old camp. We stopped several times on the way and arrived at our old camp at Young's Point on the 26th, where we remained in camp until Mar 7th and spent the time in fixing up our camp and cleaning up a little. Cold and wet weather.

Mar 8th, Our brigade got orders to get ready to go on board the boats to go up the river away to find better camping grounds, our camp here being all mud and water. We started about noon and run up about 12 miles to a place called Milligen's Bend, where we landed and went into camp. Here we remained several days and fixed up our camp etc.

Pleasant, cold and muddy.

Mar 16th, Our regiment got 3 ½ months pay today which was very gladly received.

Pleasant weather and cold.

Mar 17th + 18th, The weather is quite warm and summer like. Nothing transpired of any importance that I know of. Pleasant and warm.

Mar 19th, A boat came down the river today having on board a supply of Sanatary stores for the Wisc. soldiers in care of L.M. Gen. Treadway. The boat went on down to Young's Point where the supplys were distributed among the different regiments.

Pleasant weather.

Mar 20th, In camp. I was detailed on picket.

Pleasant and very cold.

Mar 21st, In camp. Came in from picket. The sanitary stores were distributed among us today and were very thankful for the same. Nothing of importance transpired for some days except going on picket and drilling. The weather was pleasant the most of the time and warm.

Apr 9th, All the troops at this place were reviewed by Gen. Grant today. Everything went off very nice. The weather was pleasant and cold.

Apr 12th, Our regiment received 2 months pay today and had a good old time. Pleasant weather and ~~cold.~~ warm.

Apr 14th, Our Division broke camp at 3 o'clock this afternoon and started for Carthage a point below Vicksburg. We marched out about 6 miles and went into camp on a large plantation for the night. It was very muddy + rainy.

Apr 15th, Broke camp in the morning and marched 15 miles and went into camp on a large plantation called Holmes Plantation, where we remained in camp until the 24th. During the night of the 16th we heard heavy firing in the direction of Vicksburg caused by our gunboats and transports, running the blockade. We got news the next day that 6 of our gun boats and 4 transports succeeded in running the blockade without much damage. Troops have been passing this point all day toward Carthage. Fair weather most of the time.

Apr 23rd, We lay in camp all day at Smith's plantation.

Pleasant.[3]

Apr 26th, Our brigade went on board one of the transports and run down thru a bayou into the Miss. River at a place below Carthage called Perkins plantation. We got off the boats and camped on the river bank. It rained all day and night.[4]

Apr 27th + 28th, We lay in camp at Perkins plantation all day and cooked some rations and made some preparations to attack Grand Gulf, a strongly fortified place on the Miss. River 30 miles below Vicksburg and at the mouth of the Black river. Pleasant weather and warm.

Apr 29th, Early in the morning our brigade embarked on board the transport with Gen. Austerhause's division and ran down to Grand Gulf about

---

[3] Smith's plantation refers to a plot of land owned by a local cotton planter, Pliney Smith, just north of the city of New Carthage, Louisiana, also called Carthage Landing.

[4] The Perkins plantation was a large wealthy plantation owned by the local Louisiana Judge John Perkins Jr. (1819-1885). The property was officially named the "Somerset Plantation". Perkins was a member of the Louisiana House of Representatives and the Confederate Senate. The plantation was around 17,500 acres and evaluated in 1857 to be worth upwards of $600,000 with a total of 250 slaves. At this time the plantation was occupied by Confederate forces of the 15th Louisiana Cavalry Battalion and the 1st Missouri Brigade consisting of the 1st and 4th Missouri Infantry Regiments (consolidated), the 2nd, 3rd, and 5th Missouri Infantry Regiments, and the 1st Missouri Light Artillery under the command of Confederate Colonel Francis M. Cockrell; Tucker, Phillip Thomas. "Reconnaissance in Tensas Parish, April 1863: Missouri Confederates in Louisiana." *Louisiana History: The Journal of the Louisiana Historical Association* 31, no. 2 (1990): 197.

8 miles where we lay out in the stream waiting for the gunboats to silence the rebel batteries so that we might charge the works from the water side. Our gunboats, 7 in number, commenced the attack at 10 o'clock A. M. and kept it up vigorously for 4 hrs but they could not silence 1 or 2 of their heavy guns that they had in casemate. Consequently Gen. Grant did not deem it advisable to assault the works from the river. So the troops were ordered to disembark and march down to a point about 4 miles below on the river where we arrived in the evening all night and went into camp for the night and waited the result of the blockade running our gunboats and transports were going to run the batteries in the night. Pleasant weather.[5]

Apr 30th, Last night at about 12 o'clock we were awoken by the heavy cannonading and we knew that our boats were running the blockade They succeeded in getting thru all safe 7 gun boats and 5 transports. We were got into line this morning and mustered for 2 months pay. Our troops commenced crossing the river this for noon on the gunboats and transports as fast as they could be got over. Our regiment crossed over on the gunboat Carondelet a little after noon and landed with the rest of the troops 4 miles below at a place called Bruins Burge where we all drew 5 days rations in our haversacks. The troops commenced marching out about 9 o'clock in the evening in the direction of the Magnolia Hills near Port Gibson and in the rear of Grand Gulf. Our regiment marched out at about midnight and reached Magnolia Hills about 7 in the morning, being about 15 miles. Pleasant weather.

May 1st, Morning. 7 o'clock, the fight had all ready commenced by the troops that had arrived before us. We fought the rebels all day and whipped them at every point. They were completely routed, and driven back to Port Gibson 4 miles. We took about 900 prisoners and 14 pieces of artillery. Our loss was about 1200 killed and wounded and the rebels was about 1800 killed and wounded. The rebel force was reported to be 2300 strong and under the command of Gen Bawen. This battle was fought on our side by the 13th army corps numbering about 18000 men and commanded by John A. Mcclarnand. The loss in our regiment was very small today, being only 5 wounded. We camped during the night on the battlefield and slept on our arms. Pleasant out and warm.

May 2nd, We got up this morning with the intention of renewing the fight again today but we soon found that the rebels had gathered up their traps

---

[5] Austerhause is Samuel's phonetic spelling of "Osterhaus", who was the Union Brigadier General (later Major General) in command of the Union's 9th Division of the XIII Corps in the Army of the Tennessee.

and skedaddled for parts unknown, so we got ready and started after them. We marched into Port Gibson, about 4 miles and as the rebs were in full retreat we were ordered to stop in the city and our regiment was assigned to do provost guard for a few days. The rebels also evacuated Grand Gulf yesterday and our force took possession and made it a base of supplies. Meantime the 15th corps under Gen. McPherson had gone out on another road that went out to Jackson the Capital of Miss. where a part of the rebel force had concentrated. We stayed in Port Gibson until the 5th when we marched out about 10 miles + went into camp on the bank of a bayou, where we remained until the 13th and cleaned up our clothes and drew some rations and did considerable foraging.   Pleasant.

May 13th, We broke camp this morning and started out toward Raymond where we expected to have another fight with the rebs but they did not make any stand so we marched thru the town on the 14th with out having any fight. We stopped in town a part of the day and then marched out about 2 miles and went into camp for the night. Pleasant and warm.

May 15th, We broke camp early this morning and started for Champion Hills where we met the enemy at 10 o'clock A.M. and they being in position and in line of battle we soon opened the fight with them. Our whole corps was engaged in this fight and it is said that the rebels numbered 30.000 men and 60 pieces of artillery. We fought them until dark and whipped them good. We captured about 2300 prisoners and 27 pieces of artillery. The rebel loss in killed and wounded was about 3000. We lost 800 killed and 1200 wounded. The rebs fell back during the night to Black River bridge, where they made another stand. Pleasant weather.

May 16th, Broke camp early this morning and started out for Black River bridge, a part of our forces had a fight with the rebels today and cleared them out. At Edward's Station we captured 300 or 400 prisoners and eight pieces of artillery. The loss in killed and wounded was light on both sides. We marched about 15 miles and went into camp for the night. Pleasant weather and warm.

May 17th, Gen McPherson's 17th corps had a fight at Jackson yesterday with the rebs and whipped them bad. He took a large number of prisoners and 25 pieces of artillery without much loss on either side in killed and wounded. Our corps (the 13th) moved onto Black River bridge this morning while Sherman's 15th corps went out on another to get into the rebels rear. Our corps came up to Black River and commenced the fight at about 10 o'clock A.M. We made a charge on the rebs about noon and carried their works by assault and the rebs were obliged to give way. We captured about 3000 prisoners and 17 pieces of artillery and the rebels made their tracks toward Vicksburg. The loss in killed and wounded was large on both sides. We went into camp for the night. Pleasant and warm.

May 18th, Morning in camp. We got orders to get ready to march. We moved out at noon. and crossed the river and marched 7 miles toward Vicksburg and went into camp for the night. Pleasant.

May 19th, Broke camp early this morning and started out toward Vicksburg (about 8 miles). We met the enemy about 3 miles out from the city at 2 o'clock in the afternoon and commenced skirmishing with them. We skirmished until night and drove their lines in 1 miles and went into camp for the night. Pleasant weather + warm.

May 20th, We commenced skirmishing with the rebs early this morning. We skirmished with them all day and drove them into their forts and advanced our lines up to within 700 or 800 yds of their works and went into camp under the brow of a hill out of the range of their guns where we remained in camp during the siege of Vicksburg. We lost several men today killed and wounded. Pleasant + very warm.

May 21st, Sherman opened communication with Chicksaw Bayou yesterday and we got some rations today having lived for 20 days on 7 days rations. We commenced digging rifle pits and dug all day. The weather was pleasant and very warm.

May 22nd, As the general thot that the works might be carried by an assault, we got ready to make a charge. The order was given all along the line to charge. When we all moved off at a double quick with fixed bayonets and got up to the forts without much loss on our side. We remained under the rebel forts most all day as we could not get across the ditches on the outside of the forts and could not get into the forts. We did not like to fall back in the day time as it would be the means of our losing very heavy. So we remained under the rebel forts till about sundown, as could not carry the works and then fell back with a very heavy loss on our side + with out gaining anything in return. So ended the famous charge on Vicksburg with a loss of 3000 men and nothing in return. The rebels loss was very heavy estimated at about 2000. The dead and wounded were left on the ground between the rebel lines and ours. The day was pleasant and hot.

May 23rd, We commenced the siege of Vicksburg to day in earnest by a regular siege and continued it for 40 days. We were on picket or most every day during the siege and was also under fire most every day during the siege after getting our guns in position along the lines. Pleasant.

May 29th, Our batteries opened about 500 guns on the rebel works and bombarded them furiously for about 1 hr and again at about sun set for. the same length of time. Pleasant and warm.

May 31st, There was considerable firing on both sides today but not much damage done as our batteries silenced the ~~Vi~~ rebel guns.   P l e a s a n t and warm.

June 20th, Our batteries opened at about 5 o'clock this morning and kept up a furious bombardment on the enemy works until 10 o'clock A.M. The rebels occupied their time mostly during our fire, us repairing the breaks our artillery made in their works which was quite a dangerous business for them as they lost several men killed and wounded. Our regiment was on picket today. Our artillery did some good shooting today and considerable damage to the rebel forts.   Pleasant weather.

June 23rd, The rebels opened fire on us this morning with artillery but they were soon silenced by ours as usual without any damage on our side except the killing of one man. Our brigade got orders this after noon to get ready to march with 2 days rations in haversack, as Johnson's force were threatening to attack us in the rear, but before we started, orders came from the rear that our men had driven Johnson's force back, so the order was countermanded and we remained in camp. Nothing transpired of any importance until the first of July.   Pleasant weather.

June 30th, We were got int line this morning + mustered for 2 months pay. The troops were well tired out and the weather was pleasant out and very warm.

July 1st, We got orders early this morning before light that a mine was to be exploded under one of the rebel forts and we all ordered into the rifle pits under arms to be ready for any emergency. The mine was exploded at sunrise and it took the rebels by surprise. They all raised up in their rifle pits and forts to see what was the matter when our men opened on them with artillery and musketry and kept up a steady fire for 30 minutes which was very destructive to the rebel as they lost about 2000 men. Our loss was nothing. The rebels had about 150 men blown up in the fort. Our men rushed into the fort but by some means they did not hold it. Pleasant weather and hot.

July 2nd, Our men are busily engaged preparing for some big strike. Everything remains quiet today and it is supposed that the rebs are getting short of ammunition and grub. Pleasant weather and hot.

July 3rd, This morning a flag of truce came to our lines from the rebs. The rebel General Bawen and Captain Mongomery and one orderly were admitted through our lines and were blind folded and taken to our division headquarters. They had a dispatch from Gen. Pemberton to Gen. Grant asking to be allowed to surrender the city of Vicksburg and the rebel forces holding it conditional but Gen. Grants reply was, we must have an unconditional surrender. Gen. Bawen + staff returned to Vicksburg about noon At 3 O'clock P.M. Gen. Pemberton came out under a flag of truce himself and held a long interview in which it is said he surrendered the city and forces with the understanding his troops should be paroled and allowed to go out thru our lines and go into camp, which was granted by Gen Grant for reasons of his own. Everything remained quiet along the lines. It is said that our force are going to take possession of the city tomorrow at 10 o'clock A.M. We think the 4th will be celebrated in Vicksburg. Pleasant weather and very hot.[6]

July 4th, Everything remained quiet until about 9 o'clock A.M. when a white flag was raised on all the rebel forts along the lines amidst the wildest cheering from our troops. Gen. Pemberton surrendered to us the city with 32000 prisoners, 260 pieces of artillery and all their ammunition, all in good order. Their grub was played out. After the surrender, I, with several of our boys went over to the rebel forts to see the prisoners and the forts. We went back to camp in the after noon and celebrated the 4th in the evening. Pleasant weather.

July 5th, Our division got orders to get ready to march with 2 days rations in our haversacks for Jackson where Joe Johnson has concentrated his forces numbering about 40.000 men. We immediately got ready to march without any gun sling, being nearly tired out from the effects of the long siege. We started out about noon and marched out 7 miles and went into camp for the night. Pleasant.

July 6th, Broke camp this morning and marched out to Black River about 7 miles and joined the balance of the troops that were going to Jackson. Gen. Sherman commands the expedition and Grant stays at Vicksburg. We went into camp for the night near Black river.   Pleasant and hot.

July 7th, All the force at this place broke camp this morning and started for Jackson. We marched about 16 miles and went into camp for the night at Wilson's creek. Cloudy + rainy.

---

[6] "Bawen" in this context refers to Confederate General John Stevens Bowen. Contemporary historians such as Phillip Thomas Tucker refer to Bowen as the "forgotten Stonewall of the West" for his critical role in the Western Theater. Bowen died just as his career as a military leader was gaining traction.

July 8th, Broke camp this morning and marched 16 miles and went into camp for the night near Clinton. Pleasant and warm.

July 9th, We broke camp this morning and started on thru Clinton where our cavalry met the rebs and drove them out of the place. We followed them up and overtook them 4 miles from Jackson where we commenced skirmishing with them and drove them about one miles where we got into line of battle and lay on our arms during the night. We marched 12 miles today. The loss was high on both sides.   Pleasant and hot.

July 10th, We got our breakfast early this morning and advanced on the enemy in line of battle and drove them into their works near the city. We took our position about 200 rods from the rebel works and commenced throwing up a line of rifle-pits. A continual fire was kept up all day by the artillery and sharpshooters on both sides. We lost several men killed and wounded during the day. The rebel loss was not known.   Pleasant and hot.

July 11th, Gen Sherman moved up his forces today and formed his lines from the Pearl river above the city to the river below and around the rebel works and commenced to shell the place the rebels repelled with shot and shell and we had quite a lively time all day. There was several men killed and wounded on both sides .   Pleasant weather.

July 12th, Today Gen Sherman began to press the enemy on his left and they came out of their works and charged one of our batteries (the 1st Wisc.) 20 pounders but our boys gave them such a dose of grape and canister that they went back into their works in confusion with a heavy loss in killed and wounded. Our loss was very slight. Pleasant weather and very hot.

July 13th, Our division moved a little to the right today and threw up another line of rifle pits in our front. There weas but little firing on either side today. Pleasant.

July 14th, We advanced our lines today about 500 yds and threw up a new line of rifle pits. I was out on picket today. Pleasant and very hot.

July 15th, The rebs gave our picket line a severe shelling today. I came off duty (picket) this forenoon as our lines were formed in a strip of woods, they could not see to get range on us so they did not do us much damage. A bagage of our troops on our right made a charge on the rebel works today and was repulsed with a heavy loss on our side. We lost in killed and wounded about 400 men. The rebel loss was ~~considerable~~ about 200 in all. There was considerable firing all along the lines on both sides. Pleasant and hot.

July 16ᵗʰ, The rebels returned the compliment of yesterday, today and came out of their works on our right and charged our lines but our boys repulsed them easily with a heavy loss on their side. Our loss in this affair was very small. There was very heavy firing along the lines on both sides today. The rebels gave us some lively music in the city this evening from one of their brass bands and it was believed by the most of our boys that they were preparing to evacuate the place. Our ammunition train came up today from Vicksburg and it is the intention of Gen. Sherman to make a demonstration on the place tomorrow if not a general charge. Pleasant weather and hot.

July 17ᵗʰ, Our boys awoke this morning and found that the rebs had left during the night bag and bagage for parts unknown. They destroyed what material of war they could not take along with them. They left us 6 or 8 heavy cannon. Our cavalry pressured them this morning for a few miles and captured about 1000 prisoners that had stragled behind. As soon as we got our breakfast a lot of us went over into the city to see what we could find. We got a bit of tobacco, sugar and other things, also all the peaches we wanted. The city is the finest I have seen in the South but our boys burnt a large part of it before we left. We staid around in the city till about noon and then went back to our all intrenchments and got ready to start back for Vicksburg. We got ready about 2 o'clock and marched back about 2 miles and went into camp for the night. Pleasant and hot.

July 18ᵗʰ, We remained in camp all day and went our foraging. We drew some rations today. I went out on picket. Pleasant and hot.

July 19ᵗʰ, Came in from picket, lay in camp all day and rested. Pleasant all day + hot.

July 20ᵗʰ, In camp. Washed up my clothes and lay around all day. Pleasant.

July 21ˢᵗ, Broke camp this morning and started back to Vicksburg. We marched to a place called Mississippi Springs where we went into camp for the night, it being about 12 miles. Pleasant weather and hot.

July 22ⁿᵈ, Broke camp early this morning and marched 10 miles and went into camp for the night on our old battle ground at Champion Hill. We passed thru Ray mond this after noon. It was quite a large place and a nice town. Pleasant and very hot.

July 23ʳᵈ, Broke camp early this morning and marched into our old camp in the rear of Vicksburg late in the evening, it being about 16 miles. We got all the peaches we wanted on the way back.   Pleasant.

July 24th, The 13th army corps got orders today to go into camp over on the river bank below the city so we got ready and marched up thru the city and about 1 mile below on the river and went into camp where we remained until Aug 23rd and rested up after our Spring + Summer campaign. We did not have much to do while we lay here except to drill a little and have dress parade in the evening. I went up into the city several times while laying here. The weather was pleasant most of the time.

Aug 19th, The steamboat City of Madison blew up today at the dock. She was being loaded with ammunition to go down the river. They had got her about loaded. when one of the men let a box of percusion shells drop in the hold of the boat which caused them to explode and ignite with the other ammunition which blew up the boat and killed several men. The boat was a total wreck. The weather was very hot with frequent showers.

Aug 20th, They commenced furloughing the soldiers about this time and several of the boys in our Regiment got furloughs. Pleasant.

Aug 23rd, Our brigade got orders to get ready to embark on broad the boats for New Orleans. We packed up and went on board at noon, and started on down the river. We arrived in Natchez at 4 o'clock P.M. and stopped at the landing one hour. We then went down on the river about 6 miles and tied up for the night, where we found a large melon patch and got all the melons we wanted. Pleasant.

Aug 24th, We started out about 10 o'clock A.M. and run down to the mouth of the Red River (60 miles) where we arrived in the evening and anchored out in the middle of the river over night on account of guerrillas. Pleasant + very hot.

Aug 25th, We started out down the river about the middle of the forenoon. We passed Port Hudson at noon and the City of Baton Rouge at 2 o'clock and arrived at Carton about ~~the~~ midnight and lay on the boats till morning making 170 miles run. Pleasant weather and hot.

Aug 26th, We got off the boats early this morning and marched out back at the town and went into camp. Carton is quite a large town and is situated about 7 miles above New Orleans on the river. Pleasant and warm.[7]

---

[7] "Carton" also spelled "Carlton" by Samuel refers to the historic uptown New Orleans neighborhood of Carrollton situated at the heart of the Crescent City.

Aug 27ʰ, We moved our camp today a little out back of the town and into line with the balance of the brigade in a large field. Pleasant and warm.

Aug 28ʰ, In camp. We washed up our clothes and cleaned up a little and fixed up our camp. Pleasant weather.

Aug 29ʰ, In camp. Got some ~~letters~~ mail + wrote some letters. Had dress parade in the evening. Pleasant and warm.

Aug 30, Sunday. In camp. Had inspection in the morning and dress parade in the evening. Wrote some letters during the day.   Pleasant and hot.

Aug 31ˢᵗ, In camp. We mustered today for 2 months pay. We had dress parade in the evening. Pleasant and hot.

Sept 1.2+3ʳᵈ, In camp. We had company drill in the morning and battalion drill in the forenoon and dress parade in the evening. Pleasant weather. We had all day the 3ʳᵈ to clean up our guns and traps and get ready for a grand review tomorrow. Pleasant + hot.

Sept 4ʰ, In camp. The 13ʰ Army corps was reviewed today by Gen. Grant and staff accompanied by Gen Bankes and staff. The corps made a grand appearance drawn up in line of battle by divisions. Gen. Grant was re'cd with great enthusiasm by the troops who cheered him as he passed down each of the lines and everything went off in the best of order. Gen. Grant received a severe injury by being thrown from his horse while going back from the field to the city Gen Grant was trying the speed of his horse with Gen Banks at the time of his accident. There was a very large number of people up from the city to witness the review. The weather was pleasant and very warm.

Sept 5th, We remained in camp here till the 3ʳᵈ of October and drilled a little every day. I went down to the city of New Orleans twice during the time to see the city. Our brigade went up the river to Dansflaville once during our stay here to look after some guerillas (it was 90 miles) We went out in to camp about 15 miles and did some foraging and then went back to the boats. We were done 3 days but did not find any guerillas. Pleasant + hot.[8]

Oct 2ⁿᵈ, In camp. got orders to get ready to march so we made preparations according.   Pleasant weather and hot.

Oct 3ʳᵈ, Broke camp today at Carlton and started out on our fall campaign in the Sesh country. left Carlton on boats and went down the river to New

---

[8] "Dansflaville" is likely referring to Donaldsonville, a small town just south of Baton Rouge and Carlton, Louisiana.

Orleans and landed on the opposite side of the river at Algiers. Camped in the town over night. Pleasant.[9]

Oct 4th, We lay in Algiers till 4 o'clock P.M. when we embarked on the cars for Brashere City, arrived at the city at 9 o'clock in the evening it being 85 miles. We got off the cars and went on board a boat and crossed over the bay and went into camp for the night. Pleasant.[10]

Oct 5th+6th, In camp. Fixed up our tents and cleaned up a little.   Pleasant.

Oct 7th, We broke camp this morning and left Brashere City for Opelausas. We marched 14 miles and went into camp for the night in large sugar mill. Pleasant.

Oct 8th, We broke camp and marched 18 miles and went into camp for the night. We passed thru a town today called Franklin. We were all very foot sore. Pleasant.

Oct 9th, Broke camp this morning and marched 16 miles and camped for the night. Pleasant weather.

Oct 10th, Broke camp this morning and marched 18 miles. Passed a town 3 miles out called New Iberia. Went into cam for the night 3 miles short of Vermilion. Pleasant and warm.

Oct 11th, In camp. Fixed up our tents a little and had inspection of arms + ammunition.   Pleasant weather.

Oct 12th, In camp. Nothing transpired of any account.   Pleasant.

Oct 13th, In camp. I was detailed on picket   Pleasant weather.

Oct 14th, In camp. I came in from picket. All quiet along the line. Pleasant + warm.

Oct 15th, We broke camp this morning and marched 14 miles and went into camp for night on Buzards Prairie, foot sore and well tired out. Pleasant + hot.[11]

---

[9] The period slang "Sesh" or "Secesh" was a term equivalent to "secession" or "secessionist", primarily aimed at those who aligned with the Southern Confederacy and secessionism from the United States.

[10] Brashere City refers to the city of Brashear located in modern day area of Morgan City, Louisiana.

[11] "Buzards" or Buzzards Prairie refers to the Chretien Point Plantation near modern day Sunset, Louisiana where the Battle of Buzzard's Prairie occurred on October 15, 1863 and was the precursor engagement to the more notable Battle of Grand Coteau (Battle of Bayou Bourbeux).

Oct 16th, We moved our camp a little today and fixed up our tents.   I was detailed on picket and did some foraging. Pleasant.

Oct 17th, In camp I came in from picket. All quiet along the lines. Pleasant.

Oct 18th, Sunday. In camp. Had inspection of arms and ammunition. Pleasant and very dusty and warm.

Oct 19th, Our regiment and the 34th Indiana with some cavalry was ordered out on a scout today. We met the enemy's pickets about 4 miles out and drove them back about 2 miles when we came up to their main force which was too strong for us to attack. The object of the expedition to ascertain their strength and this being accomplished we went back to camp in the evening with a loss on our side of 1 killed and 3 wounded. The enemy lost 19 killed and wounded by our artillery .   Pleasant weather.

Oct 20th, In camp. We got orders today to get ready to march tomorrow, which we did. Pleasant weather.

Oct 21st, We broke camp at 6 o'clock this morning and started out to attack the enemy. We met them about 6 miles out at 10 o'clock A.M. and charged on them and they retreated for Opelausas where they undertook to make a stand but we followed them up so close that they did not have time so they skedddled for parts unknown. We lost 1 man killed and 5 wounded and the rebels lost 35 men killed wounded and prisoners. We marched thru Opelausas and out to Bars Landing making a distance of 20 miles. I was detailed on picket thru the night. Pleasant.

Oct 22nd, In camp. I came in from picket All quiet along the lines. Pleasant.

Oct 23rd, In camp. It rained all day and very muddy.

Oct 24th, We had Regt. inspection of arms + ammunition. Pleasant weather + muddy.

Oct 25th, Sunday. In camp. Had inspection of arms this morning. Pleasant.

Oct 26th, In camp. I was detailed on picket. All quiet. Pleasant.

Oct 27th, In camp. Drilled forenoon and after noon in the manual of arms by battalion. Pleasant and muddy.

Oct 29th, In camp. It rained allday + night.

Oct 30th, In camp. Drilled forenoons and afternoons in the manual of arms by battalion. Pleasant and muddy.

Oct 31ˢᵗ, In camp. We ~~broke~~ mustered in for 2 moths pay. I was on picket. Pleasant weather.

Nov 1ˢᵗ, We broke camp this morning at 4 o'clock and marched back to Buzard's Prairie wear Carion Crow Bay on, making 20 miles and went into camp on our od camp ground. Rained fore noon. Pleasant afternoon and night.

Nov 2ⁿᵈ, In camp. The enemy attacked us today and we imeadiatly got into line of battle and went out to meet them. We had quite a sharp little fight with them and drove them back and whipped them bad. Our loss was 2 killed and 4 wounded. The rebels left 19 dead on the field. There loss in wounded was not known. We marched back into camp in the evening. Pleasant and cold.

Nov 3ʳᵈ, In camp. We were attacked again this morning by the rebs and drove them back and then went back into camp and had a state election and also drew two months pay in the forenoon. In the afternoon we were again attacked by the rebels with a large force. We went out again to fight them but they were too strong for us and after fighting them about 1 hour they drove us back about 1 mile. Our brigade numbered about 1050 men with a battery of 6 guns and 400 cavalry. The enemy number 3000 infantry and 4000 cavalry. They had us flanked on our left and rear before we commenced falling back. We formed again on the Prairie and reinforcements came up to us when we moved on to the rebs again and drove them 6 miles and whipped them bad. We held our ground until dark when we fell back to our old camp. We lost 600 men out of our brigade and one piece of our artillery. We saved all of our train and ammunition. The rebels distroyed a part of our camp equipage. Their loss was 700 killed, wounded and prisoners. We fell back during the night about 3 miles to where our troops were encamped. Pleasant + quite warm.

Nov 4ᵗʰ, In camp. I commenced cooking for the company today as our cook was taken prisoner Our Regt. lost in yesterdays fight 127 men out of 194 that was in the fight. Pleasant weather.

Nov 5ᵗʰ, Broke camp this morning and started for Vermilion (it being 12 miles) and went into camp for the night.

Rained all day.

Nov 6ᵗʰ, In camp.   Pleasant weather.

Nov 7ᵗʰ, We broke camp this morning and started back for New Iberia. We made 8 miles and went into camp for the night. Pleasant weather.

Nov 8th, Sunday. . Broke camp this morning this morning and started on for New Iberia. Arrived there in the after noon (it being 12 miles) and was assigned to do provost guard in the town and had a good time. Pleasant.

Nov 9th, In camp. Did some washing and cleaned up generally.   Pleasant.

Nov 10th, In camp. Did some washing for the boys.   Pleasant.

Nov 11th, In camp did some washing for the boys and fixed up our camp. Pleasant.

Nov 12th, In camp. I went up town today . Gen Burbridge, commanding the Post, made the citizens turn out today and did rifle pits for the defense of the town. They did not like it very much. Pleasant and cool.

Nov 13th+14th, In camp. Did some washing for the boys and went up town. Pleasant.

Nov 15th, Sunday. In camp. Had inspection of arms this morning and dress parade in the evening. We remained in camp till Dec 7th.   Pleasant.

Nov 18th, In camp. Our regiment moved out to rifle pits today and joined the bagage and was relieved from doing provost guard.   Pleasant all day.

Nov 19th, In camp.   We fixed up our camp today and polished up our guns. Pleasant.

Nov 20th, In camp. The first brigade of the 3rd division went out on th̶ a scout today and captured a 130 rebels and their horses and equipment complete with out any loss on either side. Cloudy weather.

Nov 21st, In camp. Did some washing. Nothing transpired of any importance this month out. Weather changeable.

Dec 1st, In camp. Did some washing. We remained in camp till the 7th. Pleasant.

Dec 7th, We broke camp at 10 o'clock today and started back for Brasher City. We marched 10 miles and went into camp for the night. Pleasant thru the day and rained thru the night. .

Dec 8th, We broke camp at 6 O'clock this morning and marched to Franklin, about 18 miles and went into camp for the night. Cloudy.

Dec 9th, Broke camp this morning at 6 o'clock and marched 18 miles and went into camp for the night, foot sore + tired .   Pleasant.

Dec 10th, Broke camp this morning and marched to Brasher City , about 8 miles, and went into camp for the night.   Cloudy.

Dec 11ᵗʰ, In camp. Drew some rations.   Rainy day.

Dec 12ᵗʰ, We broke camp this afternoon about 5 o'clock and crossed over the bay by boat and slept in the depot in the city over night.   Cloudy and rainy.

Dec 13ᵗʰ, We embarked on board the cars at 10 o'clock thus fore noon and started for Algiers. Arrived there at 6 o'clock in the evening. We marched out back of the town and went into camp for the night.   Pleasant.

Dec 14ᵗʰ, In camp, fixed up our tents and cleaned up.   Pleasant.

Dec 15ᵗʰ, We got 2 months pay today and had an old bender.   Pleasant all day.

De 16ᵗʰ, We remained in camp here until the 26ᵗʰ and fixed up for a winters campaign. I went over to the city once to see the sights. The weather was pleasant generaly and the boys had fine old times while in camp here.

Dec 26ᵗʰ, In camp at Algiers. We got orders this morning to get ready to go to Texas. We broke camp at 5 o'clock in the afternoon and embarked on board the ocean steamer De Malay for Texas. We left New Orleans at 10 o'clock in the evening and made 40 miles down the river and lay to on account of fog.   Cloudy.

Dec 27ᵗʰ, We hove anchor at 6 o'clock this morning and started down the river. We passed Forts Jackson and Phillips at 8 o'clock A.M. We passed out of the river into the gulf about noon and took course for Texas. A very severe thunderstorm came up about noon and lasted all day. The sea was very rough and the boys were most all sea sick. Very rough and stormy thru the night.

Dec 28ᵗʰ, Morning.   A very rough sea and pleasant over head. The boy all getting better of their sea sickness. The ship is making 10 knots an hour. A very windy night.

Dec 29ᵗʰ, Morning.   Making 10 knots an hour. We came in sight of land at noon. We run down the coast and anchored at 5 o'clock P.M. out side the bat at Mattagarda Bay, as we could not cross the bar and bay on this boat over night. A rough and stormy night.

Dec 30ᵗʰ, Morning. Pleasant over head but very rough and windy so that we can not land. A small steamer came out to us and we sent a messenger ashore at 3 o'clock P.M. WE lay at anchor all day and night. It blew a regular gale through the night. Cold and stormy at night.

Dec 31, Morning. The gale still raging and we are still laying at anchor off Mattigarda Bay. We mustered in today for 2 months pay on board the boat. Very rough and cold all day and night.

The author's 3rd Great-Grandfather, Samuel's younger brother Albert C. Burdick c.1865. *Courtesy of Daniel Scharfenberg.*

Albert Burdick c.1890s. *Courtesy of Daniel Scharfenberg.*

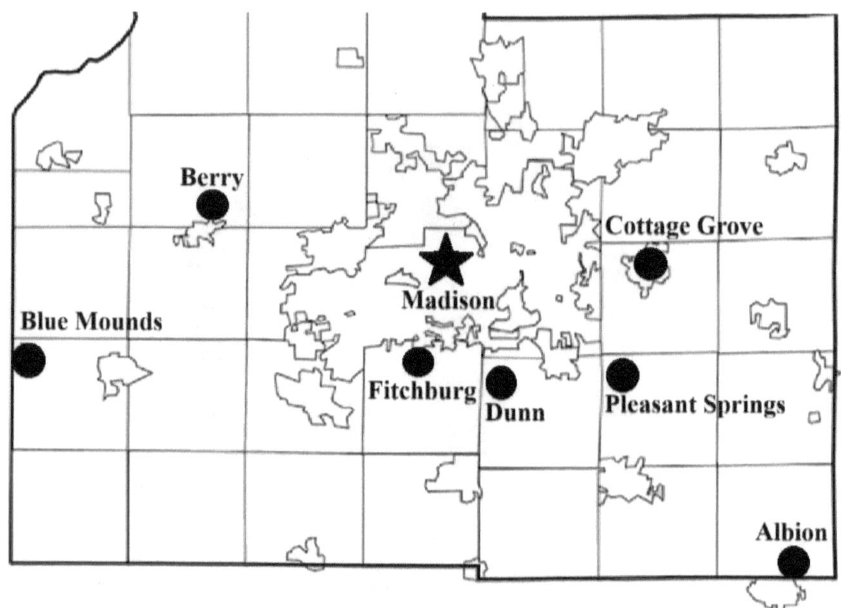

Cities and villages where Company D were primarily recruited from.
Courtesy of Daniel Scharfenberg.

Counties where the 23rd Wisconsin Volunteer Infantry Regiment was raised.
Courtesy of Daniel Scharfenberg.

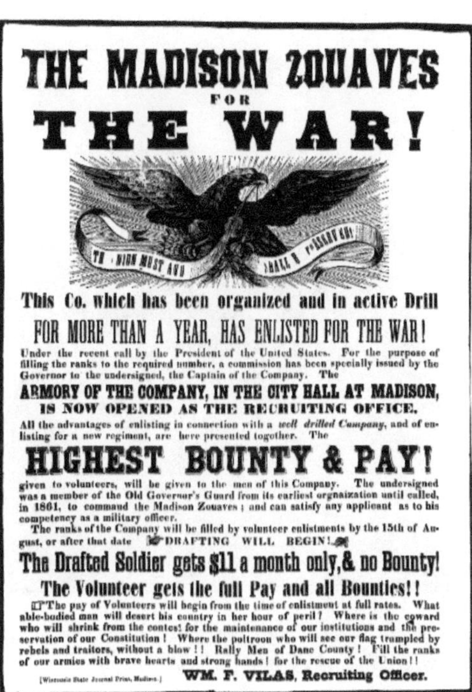

Recruitment poster for Company A of the 23rd Wisconsin, aka the "Madison Zouaves". *Courtesy of the Wisconsin Historical Society.*

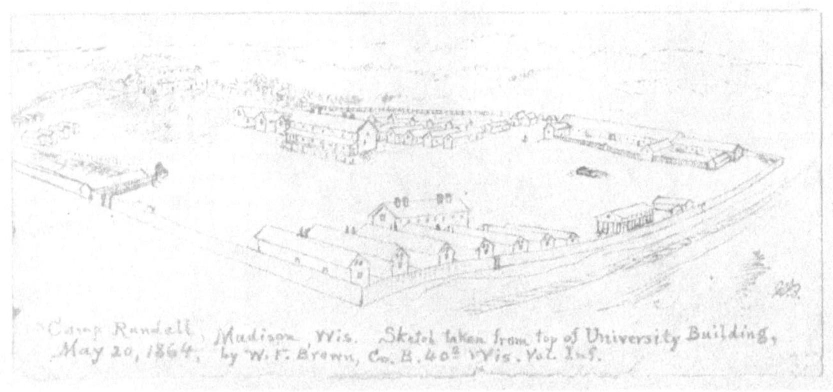

Sketch of Camp Randall from the University of Wisconsin building by William Fiske Brown of Company B, 40th Wisconsin Infantry Regiment. *Courtesy of William Fiske Brown.*

Joshua James Guppey, commander of the 23rd Wisconsin Infantry Regiment. *Courtesy of Genealogical and family history of the state of New Hampshire, Volume II.*

Lieutenant Colonel William Freeman Vilas who briefly commanded the regiment in 1863 before his resignation. *Courtesy of the Wisconsin Historical Society.*

Lieutenant Colonel Edmund Jüssen c.1885.
Courtesy of The Wisconsin Magazine of History.

Lieutenant James L. Baker c.1865.
Courtesy of Civil War Photo Sleuth.

Colonel John Leal Haynes of the 1st Texas Cavalry Regiment (Union) with his wife Angelica Irene Wells. *Courtesy of the Lawrence T. Jones III Texas Photographs, DeGolyer Library, Central University Libraries, Southern Methodist University.*

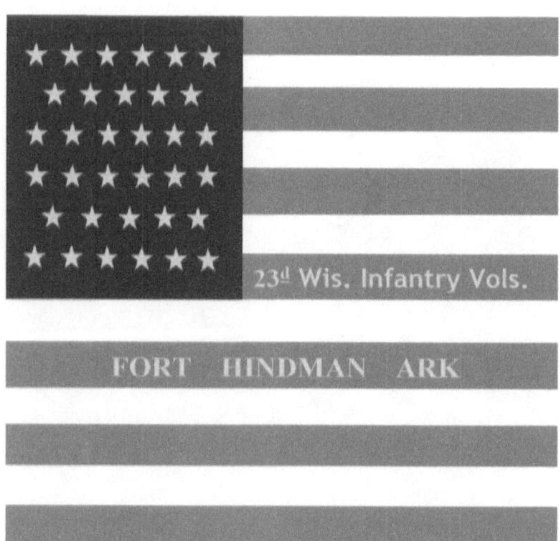

Artistic depiction of the 23rd Wisconsin's national flag which includes the battle honor "Fort Hindman" for the regiment's role in the Battle of Fort Hindman (also called the Battle of Arkansas Post). *Courtesy of Daniel Scharfenberg.*

1864

INTRODUCTION

1864 would be a trying time for Samuel and the 23rd Wisconsin. As the war took its toll on the civilian populace a new form of warfare began to be introduced to western Louisiana as the Confederate armies located in Louisiana, Arkansas, and Texas among others grew more and more isolated, this form of warfare of course was guerrilla warfare. Guerilla warfare had been a common practice in states further north of Louisiana during the American Civil War, nominally states such as Kansas and Missouri whose pre-war conflicts and border disputes spilled over into the ongoing war. With the Union Army's success at Vicksburg in 1863 many Confederate units to the west of the Mississippi were now cut off from many of the armies to the east. With the Confederacy now torn in two the Union could now impose itself upon southern states such as Louisiana and Arkansas and move freely across most rivers without opposition.

A major Union offensive was planned for 1864, this operation of course was the now infamous Red River Campaign, also called the Red River Expedition which ended in a Confederate victory. The Red River Campaign was led by Major General Nathaniel P. Banks who had previously fought in the Eastern Theater in the Shenandoah Valley campaign of 1862 against Confederate General Thomas "Stonewall" Jackson. Failing to reinforce the Union offensive against Richmond in the Peninsula Campaign, Banks was sent west and placed in charge of the Department of the Gulf. Opposing Banks was Confederate General Edmund Kirby Smith who commanded the Trans-Mississippi Department. Smith was now a seasoned commander of the Western Theater and although outnumbered proved to be a fierce defender of Western Louisiana in engagements such as the Battle of Mansfield and the Battle of Pleasant Hill. 1864 would prove to be one the toughest years of the war for Samuel and the rest of the 23rd Wisconsin due to the harshness of the Red River Campaign and the difficulty of seizing Shreveport, Western Louisiana, and Eastern Texas due to the dense woodlands and swamps that surround the Gulf of Mexico's coastal plain.

ENTRIES

Jan 1st, Morning. It was very windy, rough and cold this forenoon. In the after noon the wind went down and the steam boat Planter, came to take us ashore, everything being transported from from the De Malay to the Planter. We started for shore at 5 o'clock in the after noon and landed about 6 o'clock all safe and sound on a point of land (or a sand bar) called Decrous Point at Mattagarda Bay and went into camp for the night in the sand. Cloudy weather and quite cold.[1]

Jan 2nd, In camp. We drew rations today but had no wood to cook with, as we had to use driftwood and there was none within 7 miles of camp. Cloudy + Rainy.

Jan 3rd, We remained in camp on the sand for here till the 20th. Nothing of importance transpired while laying here. We went fishing several times and also got all the oysters we wanted. Our division was reviewed on the 13th by Gen Ransom. We got orders on the 19th to get ready to go up the bar on a 5 days scout.   Pleasant.[2]

Jan 20th, We broke camp this morning (our Brigade) and marched up the peninsula, 14 miles and went into camp for the night.   Pleasant weather.

Jan 21st, Started up the peninsula this morning, marched 18 miles and went into camp for the night. We got plenty of sweet potatoes and fresh meat. Pleasant.

Jan 22nd, We marched on up the peninsula this morning 14 miles and not finding any rebels we started back for camp, marched back 6 miles and went into the camp for the night.   Pleasant.

Jan 23rd, We started out for camp this morning it being 22 miles and went into camp for the night.

---

[1] "Decrous Point" refers to DeCrow's Point in Matagorda Bay near Texas, an area where the 23rd Wisconsin Regiment was sent while on a forward reconnaissance expedition in January of 1864.

[2] Thomas Edwin Greenfield Ransom was a General in the Union Army and commanded General McPherson's XVII Corps.

Jan 24th, We started on for camp this morning it being 22 miles. We got into camp at 5 o'clock in the afternoon, pretty well tired out and hungry as bears, with out accomplishing anything. Very pleasant weather for the winter.

Jan 25th, We remained in camp here until Feb 22nd and drilled a little during the time. We went fishing several times. The weather was pleasant most of the time.

Feb 12th, The war veterans of the 11th Wis. Regt. came to our Regt today and I left off cooking for the Co. today. Pleasant all day.

Feb 22nd, We broke camp this morning and marched down to the landing and marched on board the steamer Alliance for New Orleans. We went on board at 5 o'clock in the evening and lay on the boat over night at the dock. Pleasant weather and quite warm.

Feb 23rd, We steamed out of Mattagarda Bay at six o'clock this morning and started on our course to the mouth of the Mississippi river. We made good lead-way all day. Very pleasant all day.

Feb 24th, Morning.   Making fine head way. We arrived at the South Pass of the Mississippi river at 12 o'clock in the night and anchored o account of the fog. Pleasant

Feb 25th, Morning. We lay at anchor until 10 o'clock A.M. when the fog cleared up and we crossed the bar into the Mississippi River and went on up to New Orleans. We passed forts Phillips and Jackson at 5 o'clock P.M. and arrived at Algiers at 2 o'clock in the night. We lay on the boat until morning.   Pleasant.

Feb 26th, We unloaded our camp equipment from the boat this morning and onto the cars for Brashear City. We left Algiers at 10 o'clock A.M. on the cars and arrived at there at 5 o'clock P.M. We unloaded our camp equipment from the cars onto a boat and crossed over the bay and went into camp for the night. Pleasant all day.

Feb 27th + 28th, In camp. Washed up my clothes and fixed up our tents. Pleasant.

Feb 29th, In camp. We mustered in for 2 months pay. Pleasant weather and very muddy getting around.

Mar 1st, In camp. We remained in camp here until the 7th. We turned over our A tents the 6th and drew day tents. Pleasant and cold.

Mar 7th, We broke camp today and started our on Banks famous Red River campaign. We marched 16 miles and went into camp for the night. Pleasant all day.

Mar 8th, We broke camp this morning and marched 5 miles beyond Franklin, making 16 miles and went into camp for the night. Pleasant weather.

Mar 16th, We broke camp this morning at 6 o'clock ad marched 15 miles and went into camp for the night.   Pleasant.

Mar 17th, We broke camp this morning and marched 6 miles beyond New Iberia making 12 miles and went into camp for the night.

Mar 18th, Broke camp this morning and marched to Vermilion Bayou, about 16 miles and went into camp for the night. Pleasant.

Mar 19th. We broke camp this morning and marched to Carion Crow Bayou and went into camp on our old battle ground it being 17 miles. We took a tramp on the old battle field.   Pleasant

Mar 20th, We broke camp this morning and marched to Little Washington about 17 miles and went into camp the other side of the town.   Cloudy.

Mar 21st, We lay in camp today and rested. It rained all day and the mud was about ankle deep.

Mar 22nd, We broke camp this morning and started on for Simmesport. We marched 20 miles and went into camp for the night in a large plantation. Pleasant and cool.

Mar 23, We broke camp this morning and marched 20 miles and went into camp for the night. Pleasant.

Mar 24th, We broke camp early this morning and marched 22 miles and went into camp for the night. It rained all day and night.

Mar 25th, We broke camp early this morning and marched 12 miles and went into camp for the night. Pleasant.

Mar 26th, We broke camp at 7 o'clock this morning and marched 4 miles beyond Alexandria (making 10 miles) and went into camp for the night. Pleasant all day.

Mar 27th, In camp. We had Regt. Inspection today and signed the pay rolls for 2 months pay and washed up my clothes.   Pleasant weather and dusty.

Mar 28th, We broke camp this morning and marched 18 miles and went into camp for the night. Pleasant.

Mar 29th, We lay in camp until noon to guard the ammunition train. We started out at noon and marched 8 miles and went into camp for the night. Pleasant and muddy.

Mar 30th, We lay in camp again today until noon, guarding the train. Moved out at noon and marched 12 miles and went into camp at Cane River. Pleasant and dusty.

Mar 31st, We broke camp this morning and marched 16 miles and went into camp for the night. Pleasant.

Apr 1st, We broke camp this morning and marched 12 miles and went into camp for the night. Pleasant.

Apr 2nd, We broke camp this forenoon and marched into the town of Natchitoches and our Regiment was assigned to do provost guard in the town. Pleasant.

Apr 3rd, In camp. I was detailed today to guard the prisoners that our cavalry had captured. Pleasant.

Apr 4th, In camp. I came off guard and got some mail and wrote letters and cleaned up clothes. Pleasant + very dusty.

Apr 5th, In camp. I was detailed on guard in the town.   Pleasant.

Apr 6th, We broke camp this morning and marched 17 miles and went into camp for the night. Pleasant.

Apr 7th, We broke camp this morning and marched to Pleasant Hill, 17 miles and went into camp for the night. Pleasant and dusty.

Apr 8th, We broke camp this morning at 8 o'clock and marched about 10 miles when we met the enemy in force about nine o'clock in the forenoon, our brigade being in advance we ordered into line of battle and immediately commenced skirmishing with them and drove them 8 miles. Where we battled at Sabine Cross Roads, our division with the cavalry and artillery were got into line of battle as it was ascertained we had some up to the main force. Meantime the 19th army corps had gone into camp 7 miles in our rear. It was apparent that we were going to have a fight. Everything remained quiet till about 3 o'clock P.M. when the enemy made a furious assault on our lines and being so largely superior in numbers to us we were obliged to give away in confusion. We had 4000 infantry in our line of battle and about 1500 men. We fought them about an hour before we fell back. The rebels drove us about 4 miles when the 19th corps came up and checked the rebels for the might. Our wagon train was sent up to the front and consequently the rebels captured 100 of our wagons, also 22 pieces of our artillery and also killed, wounded and captured about 2500 of our own men, mostly from our division. Our Regt. lost 69 men, killed + wounded and captured. The rebel loss was about 6000 men. Our force fell back during the night to Pleasant Hill making a march of 36 miles. We arrived at Pleasant Hill at 2 o'clock in the morning and lay down and rested on our arms during the night. As soon as the enemy found we had gone they followed us up. The weather was pleasant and very hot. So ended the battle of Sabine Cross Roads.

Apr 9th, This morning the 19th Army corps formed in line of battle at Pleasant Hill and were immediately attacked by the enemy. The 19th Corps fought them till about noon without much success on either side but they soon began to get the worst of it and were about to give away when Gen A.J. Smith came up with 7000 reinforcements and saved the day for us. The enemy were reinforced last night about 10000 men making there force of more than 22000 men today. Our force whipped the rebels bad today and drove them over 4 miles. We captured 800 prisoners today and 8 pieces of artillery. Our loss was about 800 killed and wounded and the enemy's loss was about 2000. Gen. Smith wanted to follow up the enemy but Gen Banks ordered a retreat on account of the river being so low the boats could np get up with rations and supplies. Our loss in both days fight was about 3600, and the rebels loss was about 7000 men. Our division was sent to the rear today to guard the ammunition train and prisoners, so we were not in todays fight. Our force fell back during the night about 7 miles and rested on our arms till morning.   Pleasant weather.

Apr 10th, We started on back this forenoon at 10 o'clock, marched 10 miles. Our Regt. guarded prisoners back to Grandica, a small town on the river. Pleasant.[3]

Apr 11th, We broke camp early this morning and marched back to Grandica, about 12 miles and went into camp. Our Regt. was detailed to guard prisoners (50 men on duty each day).   Pleasant.

Apr 12th + 13th, In camp and on guard. Washed up my clothes.   Pleasant.

Apr 14th, Our Regt. moved out to the front about 1 miles and took our position in line of battle with our division and threw up rifle pits and cleared off the timber. Pleasant.

---

[3] "Grandica" refers to Grand Ecore, a small settlement north of the city of Natchitoches, Louisiana situated thirty miles west of Pleasant Hill. The river Samuel references in this entry would be the Red River, as the Red River is the only river in the vicinity of Pleasant Hill.

Apr 15th, In line of battle. We moved our line of a little today and threw up more rifle pits and cleared away the timber in our front. Pleasant weather and very dusty.

Apr 16th, We moved our camp equipage down to the rifle pits today. I was on detail.

Apr 17th, In camp. Had inspection today. All quiet along the lines. Pleasant.

Apr 18th, We had inspection today of arms and ammunition. All quiet along the lines. Pleasant.

Apr 19th, In camp. I was detailed on picket today. All quiet in front. Pleasant.

Apr 20th, In camp. I came in from picket this morning. We got orders this forenoon to be ready to march at a moments notice with 2 days cooked rations in our haversacks. Gen. Smith's forces moved out 4 miles to Natchitoches today. Pleasant and very dusty.

Apr 21st, In camp. We drew 5 days rations + got ready to move. We broke camp at 5 o'clock in the afternoon and moved out into the road and lay around in the dirt till 4 o'clock in the morning. Pleasant.

Apr 22nd, 4 o'clock in the morning. We started out and marched all day and part of the night and lay on our arms till morning. We made 35 miles. Cloudy.

Apr 23rd, We started out at 4 o'clock this morning and marched to Cane River where we met the enemy in our front in force on the opposite side of the river. Our artillery commenced shelling them across the river about 2 miles and forded the stream and around on the enemys flank when they had to skedaddle. We captured about 400 0f their horses as they were dismounted and about 30 prisoners. Our loss in killed and wounded was about 150 and the rebels was about the same. Gen Smiths force brot up the rear all the way there and had a fight everyday as the rebels followed us up. We had a fight with them today and whipped them hansomley. Pleasant weather. We remained in camp over night at Cane River. ~~Pl~~

Apr 24th, We broke camp about 10 o'clock this forenoon and marched 22 miles and went into camp for the night. Smith's forces had a fight in the rear today and cleared out the rebs as usual Pleasant and dusty.

Apr 25th, We broke camp this morning at 6 o'clock and marched 18 miles and went into camp 3 miles from Alexandria. Smith's forces had 2 fights in the rear today and whipped the rebs both times.   Pleasant.

Apr 26th, In camp at Alexandria. Washed up my clothes, drew some new clothing and got some mail.   Pleasant.

Apr 27th, In camp. Drew new day tents today. Wrote some letters and fixed up our camp. A Brigade of the 1st division came up to us today and also the 1st Wis. Battery. Our brigade all turned out to cheer McClernand who came along with them.

Apr 28th, We had brigade inspection of arms and cleaned up our camp this forenoon. We got orders to fall into line of battle at noon as the enemy was advancing on us in large force. We moved out in line of battle to meet them. We threw out a line of skirmishers which commenced skirmishing with the rebels. We skirmished with them for about 2 hrs when we fell back one mile and got into line of battle and waited for them to attack us but they did not see fit to do so. We lay in line of battle all night and the rebels fell back during the night.   Pleasant.

Apr 29, Morning. All quiet along the lines. We moved back about 1 mile this morning and got into line of battle and threw up rifle pits.   Pleasant.

Apr 30, In camp near Alexandria. We mustered in today for 2 months pay and I wrote some letters home. All quiet along the line.   Pleasant.

May 1st, In camp. We got 4 months pay today. I went down town and had a good old time. Pleasant.

May 2nd, In camp. We drew some new clothing this forenoon. We got orders at noon to be ready to move out at 1 o'clock with 2 days rations in haversack. We went out 3 miles and met the enemy and commenced skirmishing with their advance. We drove them back 2 miles and skirmished with them till night. We fell back 1 mile and lay in line of battle all night.

Pleasant all day.

May 3rd, In line of battle all day. All quiet along the line.   Pleasant.

May 4th, In line of battle. All quiet. We had inspection of arms and ammunition this fore noon. We moved out at noon to support our picket line while they advanced their line. We had a light skirmish with a squad of the rebel cavalry and then went back to camp. Pleasant weather.

May 5th, In camp. I was detailed on picket. The 1st + 3rd divisions of our corps moved out today to feel the strength of the enemy's position. They had a brief skirmish with them and drove them 7 miles with a slight loss on our side, the rebels loss not known. Our forces cam back to camp this evening. Pleasant.

May 6th, In camp. I came in from picket this morning. All quiet along the line. Pleasant weather.

May 7ᵗʰ, We lay in line of battle until the 13ᵗʰ without any molestation from the enemy. We had inspection of arms during this time and did some foraging. We were 6 miles out from Alexandria.

Pleasant weather generaly.

May 13ᵗʰ, In line of battle. We fell back to Alexandria this evening and went into camp for the night. Pleasant.

May 14ᵗʰ, We marched at 5 o'clock this morning from Alexandria. We had not gone far when we came up to the enemy. We skirmished with them awhile and drove them off without much loss on either side. We then went on a march 15 miles and went into camp for the night. Pleasant.

May 15ᵗʰ, Morning. The enemy followed us up so sloe that we had to lay in line of battle all day and skirmished with them to give our train a chance to get ahead. We marched at 5 o'clock in the afternoon and marched 15 miles and rested on our arms until morning.

Pleasant and warm.

May 16ᵗʰ, We marched at 6 o'clock this morning. Our division and 1 Brigade of cavalry brot up the rear from Alexandria threw to Yellow Bayou. Our forces had a fight on a head today and whipped the rebs with quite a loss on both sides. We marched 16 miles and lay in line of battle threw the night. Pleasant.

May 17ᵗʰ, We marched at 6 o'clock this morning and the enemy attacked us today on the march and was whipped as usual. We marched 15 miles and went into camp for the night. Pleasant.

May 18ᵗʰ, In camp. Our division remained in camp today and rested. Gen. Smith's forces went to the rear today and had a fight with the rebs. He whipped them bad and drove them 4 miles. He took 400 prisoners. Our loss was about 150 killed and wounded. The enemy loss was about 500 men. Pleasant.

May 19ᵗʰ, We remained in camp until noon when our corps moved out to attack the enemy on their right flank while Smith's forces attacked them in front but the enemy saw the move and made a hasty retreat, so we march back to camp, it being 7 miles.   Pleasant.

May 20ᵗʰ, We remained in camp till noon and then marched out to the Atchafalaya River (1 mile) and cross over on a pontoon made of steamboats. We stacked arms and lay around till 5 o'clock in the afternoon when we marched down to the Red River and went into camp for the night, making six miles march.

Pleasant.

May 21ˢᵗ, We broke camp this morning and marched down to the mouth of the Red River and down to Mississippi 4 miles where we halted and drew days rations and marched back to the mouth of the Red River and went into camp for the night, making a march of 16 miles.   Pleasant.

May 22ⁿᵈ, We broke camp this morning at 5 o'clock and marched down to the river to Marganasia about 20 miles and went into camp for the night. So ended the Red River campaign with a loss to us of about 125 wagons, 27 pieces of artillery and about 5000 men and a loss to the enemy of 33 pieces of artillery and about 9000 men. We traveled during the campaign 570 miles. Pleasant.[4]

May 23ʳᵈ. In camp. We cleaned up today and had brigade inspection. I was out on a detail all night unloading boat.

Pleasant and warm.

May 24ᵗʰ, In camp. I came in from detail this morning and washed my clothes got orders at 5 o'clock in after noon to pack up and get ready to go on board the boat for down the river. We went on board the Kate Dale at 6 o'clock and lay on her over night at the landing.

May 25, We got of the Kate Dale again this morning and lay in camp all day. We went on board of her again this afternoon at 5 o'clock and started down the river for Baton Rouge. We arrived at 10 o'clock in the evening. I was on detail to unload the boat. Worked all night and was about used up. Pleasant.

May 26, We fell into line this morning and marched out back of the city one mile and went into camp.   Pleasant + warm.

May 27, In camp. I was detailed with 60 men from our Regt. to do provost guard in the city and took up our quarters in a house. We remained here and did provost guard until the Regt. left this place and had a first rate good time. Our Regt. remained in Baton Rouge until July 7ᵗʰ.   Pleasant.

June 6ᵗʰ, I went up to the Regt. today. The 11ᵗʰ Wis boys were ordered to join their Regt. today and left us.   Pleasant.

---

[4] "Margansia" or "Morganzia" is Samuel's own phonetic spelling of the small city of Morganza, Louisiana, near the Mississippi River, located in Pointe Coupee Parish. Morganza was used as a Union encampment for much of the later war era from 1864-1865.

June 14th, I went up to the Regt today and was paid 2 months pay. Pleasant.

June 15th, I got a furlough. Started today to go home. Sent it to New Orleans to be approved at head quarters. It was pleasant weather all day.

June 16th, I went out back today to see the troops reviewed by Gen. Sickles. Everything went off nice.   Pleasant.[5]

June 28, I was detailed with 5 men and a Sergeant to guard some prisoners down. We were gone to New Orleans 6 days and had a first rate good time in the city. We got back to Baton Rouge on the 29th at night.   Pleasant.

June 30th, I went up to the Regt today and mustered in for 2 months pay. The men drilled a little at the Regt most every day.   Pleasant.

July 4th, There was no celebration here today of any account.   Pleasant.

July 7th, The Regt got orders today to pack up and be ready to board the boat for New Orleans at moments notice. Cloudy and rainy.

July 8th, We were relieved from provost guard this morning by the 19th Ky. Regt. and went down to the boats and was detailed to help load the boats with our Regt trapes. We got on board the boats and left Baton Rouge at 11 o'clock A.M for New Orleans. We arrived there at 9 o'clock in the evening and lay on the boat over night . Cloudy and rainy.

July 9th, We crossed over the river this morning to Algiers and unloaded the boats and went into camp out back of the town.

Cloudy.

July 10, Sunday. In camp. We had inspection of arms in the morning and dress parade in the evening. Cloudy + rainy.

July 11, In camp at Algiers. We remained in camp here until the 25th. We drilled most every day. I went out back blackberrying several times. The weather was cloudy and rainy.

July 18th, We turned over our old guns today and drew new ones. Pleasant forenoon, rained after noon.[6]

---

[5] Daniel Edgar Sickles was a General in the Union Army who is most well-known for his role at the Battle of Gettysburg. Sickles is considered by many to be a "political general", someone who lacked in a military background and received a political appointment as an officer.

[6] Muskets during the Civil War were issued to Union and Confederate soldiers throughout the war on several occasions, mainly as needed or when a musket needed repair services or exchange due to the rough conditions of campaigning. The 23rd Wisconsin was overwhelmingly armed with the English imported Pattern 1853 Enfield rifled musket.

July 22ⁿᵈ, I was promoted to Corporal on dress parade this afternoon. Pleasant.

July 25ᵗʰ, In camp at Algiers. We got orders this morning at 9 o'clock to be ready to go on board the boat at day light in the morning.   Pleasant.

July 26, We broke camp at 4 o'clock this morning and went on board the steamboat Gen Rogers. We started out from New Orleans at 6 o'clock for Morganzia. We passed Baton Rouge at 7 o'clock in the evening and arrived at Morganzia at 5 o'clock in the morning. Cloudy.

July 27ᵗʰ, We got off the boat this morning and unloaded the boats and went into camp on the river bank and fixed up camp. Pleasant forenoon + rained afternoon.

July 28, In camp. I washed up my clothes today and cleaned up generaly. ~~Our~~ A brigade of our division went out on a scut today to Atchafalaya 16 miles and had a fight with a the rebs and whipped them bad and drove them away. Our loss was 3 killed and 2 wounded. The enemy was not known. They returned to camp this evening. I went down to the 4ᵗʰ Wis cavalry boys that I was acquainted with.   Rained all day and night.

June 29ᵗʰ, We remained in camp at Morganzia till the 20ᵗʰ of Aug. and drilled some most every day. The weather was very warm while we lay here. Pleasant as a general thing.

Aug 10ᵗʰ, Our cavalry had a little fight out back 7 miles today. They captured 15 rebs with a slight loss on our side. The rebel's loss not known.   Pleasant.

Aug 17ᵗʰ, Our Regt had inspection today in heavy marching order.   Pleasant.[7]

Aug 20ᵗʰ, In camp at Morganzia. We got orders this afternoon to get ready to go on board the boat for New Orleans. We went on board the steam boat Silver Wave at 10 o'clock in the evening and started for New Orleans. We made 35 miles down the river and anchored for the night on account of fog. Rained thru the day and night.

Aug 21ˢᵗ, Sunday morning. We started on down the river at 7 o'clock. We arrived at Baton Rouge at 9 o'clock in the forenoon and stopped for 30 minutes and then went on and arrived at New Orleans at 9 o'clock in the evening and lay on the boat over night.   Pleasant.

---

[7] "Heavy Marching Order" according to period drill manuals meant to carry all possessions of the soldier on their persons. It was typically reserved exclusively for extended marches and was extremely uncomfortably.

Aug 22, We steamed up this morning and run down along side of the Ocean Steam boat Calanba and embarked on her and with the 161st New York inft. for Mobile Bay. We left the dock at New Orleans at noon and

Aug 23rd, We arrived of the bar at Mobile Bay this morning at 7 o'clock and cast anchor. We sent a small boat anchor for orders at Fort Morgan had surrendered this morning, we got order to run in across the bar which we did at 10 o'clock and anchored near Fort Morgan. The fleet was firing a salute of 100 guns as we went into the bay. Our fleet and land forces opened on for Morgan yesterday morning early and kept up a continual bombardment until this morning at 6 o'clock when the fort was compelled to surrender unconditional. Our forces captured 580 prisoners including the commanding Gen. Page and 630 heavy guns and all of their small arms + ammunition and also 6 mos rations for the garrison. We got off the Calanby onto the transport W.W. Thomas at 10 o'clock in the morning in the evening and run up called Pilot town 3 miles above Fort Morgan and lay on the boat over night.

Aug 24, We got off the boat this morning and went into camp on the beach. Pleasant.

Aug 25, In camp. Our fleet and land force had previously captured fort Ganes with its garrison of 600 men, also 43 heavy guns and six months rations and all their equipment and ammunition all in good order. The rebs blew up Fort Powel when they first saw our fleet. It mounted 19 guns, the most of which was saved by our men. Our fleet also captured the rebels from Tennessee and 3 or 4 other gun boats and transports. We got orders to be ready to march with 2 days cooked rations. We broke camp at noon and went on board the N. W. Thomas with our troops and crossed over the bay on the South side and 8 miles march from Fort Ganes at a point called Ceder Point where the rebs had a fort and some breast works. The rebs saw us a coming and skedaddled up the point. We made a landing after considerable trouble and formed in line of battle and marched up the peninsula about 3 miles where we came to a Bayou where the rebs had destroyed the bridge so we had to go back. We did not see any rebs as they fell back to a strip of woods about 2 miles. We selected out ground and went into camp and threw up some rifle pits We remained in camp here till Sept 1st. We got all the oysters that we wanted while lying here. The misquitos were so thick here they tormented us almost to death. Pleasant.[8]

---

[8] "Ganes" is Samuel's phonetic spelling of Gaines. Fort Gaines was a fortress on Dauphin Island in the Gulf of Mexico. Likewise, "Powel" is the phonetic spelling of Powell, Fort Powell being a fortress located on a shell island in Grant's Pass in Mobile Bay, Alabama.

Aug 31, In camp at Cedar Point. We mustered in for 2 months pay. Pleasant.

Sept 1, In camp. We got orders today to be ready with 2 days cooked rations to go to New Orleans.   Pleasant.

 Sept 2, We loaded our camp equipment on to the N. W. Thomas this morning and the rest went on board at 10 O clock A.M. with the 96th Ohio inft and the 17th Ohio battery.   We fooled around all day and tonight trying to get our artillery on board.   Pleasant.

Sept 3rd, Morning. Still trying to get our artillery on board. We succeeded in getting our artillery on board the boat at 10 O'clock in the forenoon and was getting ready to pull out from the wharf which a gale came up so fierce that we had to lay at the dock for 2 hrs when the gale subsided and we started out and run over to fort Morgan. We landed at Fort Morgan at 1 o'clock to get off the 17th Ohio Battery which was not going along with us. We went ashore and went up into the fort and after having a good look at it inside and out we went back to the boat and it started out over across to Fort Ganes on the opposite side of the channel. We lay at the dock here over night and went up into the front and all around it. We lay on the boat over night.   Pleasant for noon, rained afternoon.

Sept 4th, Sunday. We left Fort Ganes at 4 o'clock in the morning and run over to Fort Powel and got aground where we remained four hours before we could get off. Fort Powel was situated on an island that lay between Cedar Point and Fort Ganes. It was built for the purpose of guarding what was called Grant's Pass into Mobile Bay. We got off the ground about 10 oclock in the fore noon and started for New Orleans. We passed Ship Island about 4 oclock in the afternoon. The pilot lost his reckinging in the night and run on the sand bar and got stuck where we remained until morning.   Pleasant.

Sept 5th, We got off the sand bar this morning all right and started on our way. We went in at the East Pass of the Miss. river at noon, past Fort Phillips and Jackson at four o'clock in the afternoon and arrived at New Orleans at 12 O'clock at night and lay on the boat at the dock until morning.   Very pleasant.

Sept 6th, We got up the steam this morning and run up to the upper end of the city to coal up the boat. We lay at the bank all day and went ashore and drew and cooked rations. We started out at 10 oclock in the evening and run all night.   Pleasant.

Sept 7, We passed Donaldsonville at 10 o'clock this morning and arrived at Baton Rouge at 4 o'clock in the afternoon. We stopped at the landing and got aground and was one hour getting off. We passed Fort Hutson at 8 in the evening and arrived at Morgantia at 2 o'clock in the night. We lay on the boats till morning. Pleasant.

Sept 8, We unloaded our camp equipage from the boat this morning and went into camp on the river bank. Pleasant.

Sept 9th, In camp at Morgansia, We remained in camp here until Oct 1st without much excitement. We drilled some most every day and went out on some scouts. We had quite an easy time. It was pleasant weather generally and very warm. We fixed up our quarters in good shape.

Sept 16, The rebs made a dash on a squad of our cavalry that was out on a scout and captured about 30 of them, so our brigade was ordered to be ready to march with 2 days rations at sunset to go out and support the cavalry. We marched at at 6 in the evening. We met up with the Miss. river about 17 miles and lay down and rested until morning.   Pleasant.

Sept 17th, We got out breakfast and marched at 6 oclock out toward the Atchafalaya river we went about 7 miles and met the cavalry coming back, having driven the rebs across the river, so we halted and was recalled back to camp. We did not get any chance at the rebs but we did considerable foraging on the way back. We rested 2 hrs and got our dinner and then started back for camp. We made 16 miles back (23 miles in all) and went into camp for the night .   Pleasant.

Sept 18th, We broke camp this morning and marched into camp ~~this mo,~~ about 8 miles, got into our camp at 8 o'clock in the forenoon.

Pleasant all day.

Sept 20, In camp. We cleaned up our clothes this forenoon and had battalion drill in the afternoon. We got orders this morning to get ready to march at midnight with 5 days rations. We got ready and marched ~~into~~ at 11 o'clock in the night, marched all night . Pleasant.

Sept 21, We arrived at Center Port on the Atchafalaya river this morning at 5 o'clock, being about 16 miles and went into camp. We threw up some works for our protection. I was guard threw the night. Pleasant.

Sept 22nd, There not being any rebs around we allowed to forage all we wanted to. We got all the sweet potatoes chicken and fresh meat that we wanted. We got orders to be ready to go back to camp int the morning at 8 o'clock. Pleasant.

Sept 23, We broke camp this morning at 8 oclock and started back from Morgansia without seeing any rebs. We arrived at our old camp on the bank of the Miss. river at 4 oclock in the afternoon all night. It rained in the after noon.

Sep 24, In camp. Nothing of importance transpired this month out. Pleasant.

Sept 25, Sunday. In camp. We had Regt. inspection in the fore noon and went down to the 23rd Iowa Regt. visiting in the afternoon. Pleasant.

Oct 1st, In camp. We got orders to clean up our guns and traps ready for Regt. inspection in heavy marching order tomorrow. Pleasant.

Oct 2nd, Sunday in camp. We had inspection this morning. We got orders at sunset to get ready to march at 5 o'clock in the morning on the boats with 5 days rations. Pleasant thru the day, rained in the night.

Oct 3rd, We got ready this morning and started down to the boats at 6 oclock to embark with the balance of the expedition composed of 4 pieces of the 4th Mass. Battery, 200 of the 1st Texas Cavalry 300 of the 1st L.A. Cavalry and the 1st L.A. Infantry, 161st N.Y. Infantry and our Regt. all under the command of Col. J.J. Guppy of our Regt. We got everything on board the boats and started down the river at 9 o'clock in the forenoon. We run down about 12 miles to a place called Bayou Sara. We landed at 10 o'clock and imediately unloading unloaded the boat and marched out back of the town and got int line of battle and stacked ~~their~~ our arms to get our dinner. The rebel pickets were in the town when we landed but they left as soon as they ascertained about what our force was but some of them were a little too late as our cavalry got off and put out after them and got 3 of them before they could get away. The rebs had been firing on our boats at this place, a few days before which caused the expedition. After dinner we planted our artillery in position and sent out pickets and remained here until morning. The weather was cloudy and rainy.

Oct 4th, We got our breakfast at 5 o'clock this morning. When our Regt with 2 pieces pf artillery and 200 Texas cavalry were ordered to get ready to march out to Little Jackson L.A. (being 14 miles). We started at 7 o'clock A.M. under command of Col. Hanes of the 1st Texas Cavalry while the balance of the expedition remained back to guard the town and reconoitre on the roads. We marched out there without meeting any opposition from enemy. We arrived at Little Jackson at 1 o'clock P.M. and our Regt stacked arms in the streets of the town while the cavalry went out beyond the town to see what they could find and when they soon ascertained that the rebs were in force but as the Col. had the particular orders not to bring on a fight we remained in the town 2 hrs and then started back. We marched back 2 miles and crossed a creek + got a good position and went into camp for the night. The rebels followed us back to the creek where our artillery gave them a few shells and they remained quiet thru the night with the exception of firing a few musket shots at our pickets during the night. I was on picket and my post was fired on twice thru the night.   Pleasant weather and very cold.[9]

Oct 5th, We got our breakfast at 5 o'clock this morning and had just got ready to start back for the boats when the enemy, (having been reinforced during the night) opened on our camp with 4 pieces of artillery, Our 2 pieces of artillery still being in position, immediately replied to them with shell and soon silenced the enemies guns. They threw 5 or 6 shells in our camp and killed 1 man and wounded 2 more in our Regt. but the Col. having strict orders not to bring on a fight after silencing the enemies artillery and driving them back, we started for the river but the determination of making us fight We had got about half way back to the river (it being a good place for them to bring on a fight and flank us) they charged on our rearguard which brot us to a halt. We immediately got in line of battle and got our artillery into position when the fight commenced in good earnest between the artillery and the skirmishers on both sides. We fought them for one hr when the rebs getting rather the worst of it they ceased firing and we withdrew from the field, taking our dead and wounded along with us. During the fight the rebels charged on our artillery but getting too much grape and canister they skedadled back again to the rear under a most destructive fire of our artillery. We had 1 man killed and 3 wounded in our Regt in the fight As the rebs did not see fit to attack us again while lying in position here we started on toward the river but the rebs still continued to followed us up. We marched back about 3 miles and the balance of forces coming up to reinforce us so we made a stand + wanted to attack but they did no come very close so we gave them a good shelling and then marched back to the river. We got into line out back of the town and stacked arms when we got our dinner and remained for 2 hrs. We then marched down to the river and got on the boats at 4 o'clock P.M. A few of the rebels came down in to the town while we were loading on board the boats but we happened to have a gun boat laying close by which sent a few shells over into the back part of the town and they did not show themselves any more. We started out from the landing at 5 o'clock P.M. and arrived at Morgansia at 6 o'clock in the evening all right. The object of the expedition being accomplished which was for us to go out

---

[9] Samuel spells "Hanes" phonetically, referring to Colonel John L. Haynes of the 1st Texas Cavalry of the Union Army, not to be confused with the 1st Texas Cavalry (Buchel's/ Yager's 1st Mounted Rifles) of the Confederacy (see image of Colonel Hayes in the middle of book).

to Little Jackson and draw the rebs in as wear the river as we could from that place and Clinton. While a force of our men went out from Baton Rouge to get in their, which I suppose they did, as we accomplished our part of the program. We lost 25 men killed and wounded in all. We had two killed and 6 wounded in our Regt. and the rest from the cavalry. The enemy's loss was not known but it was supposed to be large as our artillery did good execution their loss was probably 80 or 100 men.

Pleasant and cold.

Oct 6th, In camp. Nothing of importance transpired until the 10th inst. Weather Pleasant generaly.

Oct 10th, We got orders this morning to clean up + get ready for general inspection at 2 oclock in the after noon, but we had to postpone it as we got orders to get ready to go on board that boats on an expedition at 4 o'clock in the after noon. The expedition was composed of 5 Regt of Infantry, 2 of Cavalry and 4 pieces of artillery all under the command of the Col. of the 24th Iowa Inft. We started from Morgansia at about 4 oclock in the afternoon and run up the river about 40 miles when it was said the rebs had been crossing cattle over the river. We landed about 11 o'clock in the night on the east side of the river and tied up for the night. Pleasant all day and night.

Oct 11th, We steamed up early this morning and run over across the river to the L.A. side when our cavalry soon landed and started out in the direction of the Red River to ascertain whether there was any enemy in that vicinity. Meantime our Infantry got off the boat and lay on the river bank in line. Our Cavalry came back about night after going out 12 miles without finding any enemy. They then went up the river about 3 miles and camped on the bank of the river for the night. The infantry lay at the boats over night. Pleasant all day.

Oct 12th, Our Cavalry moved out back into the country early this morning and our Regt with 2 others and 1 piece of artillery was ordered to follow them. We started out at 10 o'clock in the forenoon and marched out 15 miles and our cavalry went out 10 miles by and and not finding any enemy, we started back for camp, marched back 6 miles and went into camp for the night. Pleasant.

Oct 13th, During last night the boat came up with the balance of our brigade from Morgansia, having been ordered to report at White River and our stuff was unloaded from the Illinois onto the marine boat Baltic. We started back for the boats this morning and arrived there about 10 oclock in the fore noon without seeing any rebs. Our Regt. embarked on board the

Baltic for White River and started on the river at 11 o'clock A.M. We arrived at Natchez at 4 o'clock P.M. and stopped for 2 hrs when we took a small boat into Vicksburg. Soon after we started out the small boat got under the wheels of our boat and was stopped and sank but no lives lost that I know of. We run all night. Pleasant.

Oct 14, We passed Grand Gulf at 4 o'clock this morning and arrived at Vicksburg at noon when we landed before the city and cleaned off the boat and drew and cooked rations. We lay on shore 8 hrs. We went on board the boat again at 8 oclock in the evening, and started on up the river. We run all night.   Pleas ant.

Oct 15, We passed Lake Providence early this morning and run all day. We laid up at 9 o'clock in the evening 30 miles below White River on account of the fog.

Cloudy and rainy

Oct 16, We started on up the river at sunrise and arrived at the mouth of the White river at 2 oclock P.M. when we landed and went into camp. Pleasant + cold.

Oct 17th, In camp. We tried fixed up our huts and washed up our clothes. Pleasant.

Oct 18th, We got orders about 3 oclock this morning to go on board the boats to go up White River We packed up and went on board the Ell Word and got all ready to start out at 7 oclock when a boat came down the river with an order for all the troops that could be spared at this place to be out up to Memphus as the rebels were threatening to attack that place in a large force so our orders were counter marched and our bagage was ordered to go again to camp and all the other troops were ordered to Memphis. They left here about 10 oclock A.M. and we went into camp again and fixed up our tents and drew some more clothing.   Pleasant.

Oct 19th, Our brigade with the expedition of our Regt, was ordered this morning to go on board the boats for Memphis. They left here about 8 oclock in the evening and our Regt was left to guard the camp.   Pleasant.

Oct 20th, There were men detailed from our Regt. today for head quarters and provost guard. We remained in camp here and did picket and guard duty the month out. Pleasant and cool.

Oct 21ˢᵗ, We had Regimental inspection in heavy marching order today by the district inspector and was complimented by him very highly for our good appearance Pleasant.

Oct 31ˢᵗ, In camp at White River. Was mustered today for 2 months pay. Pleasant.

Nov 1ˢᵗ, In camp at White River. We got orders to be ready to have here tomorrow

Cloudy and wet.

Nov 2ⁿᵈ, In camp. We got orders today at noon to pack up and get ready to go on board the boat for Helena Ark. We went on board the Thomas E Tint. 4 o'clock P.M. We lay at the landing over night on the boas . Cloudy and cold.

Nov 3ʳᵈ, We started out up the river this morning at 6 o'clock. We made slow head way up the river and arrived at Helena at 9 o'clock in the evening. We lay on the boat over night. Cloudy, cold and wet.

Nov 4ᵗʰ, We got off the boat this morning at 6 oclock and stacked arms on the river bank. and waited for orders. Our Col. soon came back from the Gen. headquarters with the news that the 6ᵗʰ Minn. Infy. were going away and we were to take their place. They had just got their winter quarters done so we moved into them at noon and took possession. Helena is quite a large town but it shows the effects of the war like all other towns on the river. We remained in camp here and did garrison duty until Fed 20ᵗʰ 1865 We fixed up our quarters today and fixed up our camp.

Nov 8ᵗʰ, Today was the election   . I voted for Abraham Lincoln for Pres. and Andrew Jachson for Vice Pres. It rained all day and night.

Nov 10ᵗʰ, I was detailed with 20 men to go down the river 8 miles in a boat and bring in some niggers and cotton. We started down the river at 10 oclock in the forenoon with 25 cavalry on the steam boat Dane. I was detailed on picket with 8 men as soon as the boat landed while the balance of the detail loaded cotton on the boats and the cavalry went back in the county to see what they could find. The cavalry came in at dark with out finding anything. We fooled around until 2 oclock in the morning getting cotton and niggers on the boat when we started back for camp. When we arrived at 4 in the morning and turned in for some rest and sleep. Pleasant all day and cold.

Nov 23, We had Regt inspection today in heavy marching order by the district inspector and was complimented for our good appearance . Pleasant and cold.

Nov 24, We had a Thanksgiving meeting today Nothing of importance happened this month out. The weather was very wet and quite cold about this time.

Dec 3rd, I got a box of good things today from home.   Pleasant.

Dec 9th, We had about 1 inch of snow today, the first we have had this winter and it went as quick as it come

Dec 15, Maj. Gen. M. A. Gillman revisited this place today to inspect the fortifications of the town and was saluted with 13 guns. He reviewed the troops also. Capt Sumner of our company started for home today on a leave of absence, being sick.

Cloudy and rainy.[10]

Dec 24th, We fired a salute from the forts today of 100 guns in the honor of Gen. Thomas great victory up in the Tennessee over Hoods army. Pleasant and quite cold.[11]

Dec 29th, Maj. Green came back to the Regt today after being sick for 2 months.

---

[10] Gilman Marston was a Union General originally from New Hampshire noted for his participation in the Peninsula Campaign, the Second Battle of Bull Run, and the Battle of Fredericksburg.

[11] George Henry Thomas was a southern Union General who was made famous during the Battle of Chickamauga. Thomas was dubbed "The Rock of Chickamauga" for his stubborn resistance against Confederate troops during the ferocious battle. Meanwhile, John Bell Hood was a Confederate General who commanded the "Texas Brigade" or "Hood's Brigade" in many battles of the Eastern Theater and later fought in the Western Theater during the Franklin–Nashville campaign.

1865

INTRODUCTION

1865 would be Samuel and the regiment's final year of the war. With the campaigns of the Union's XIII Corps taking its toll on the states of Louisiana, Mississippi, Texas, and Alabama, the war was soon to be over. 1865 introduces several men to the diary text that have little introduction or follow up by Samuel in his own diary entries. The first person of interest is Elisha Peckham Clark or "E.P. Clark". Elisha Peckham Clark (1833 – 1904), was a 1st Lieutenant and the assistant regimental surgeon of the 31st Massachusetts Infantry Regiment.[1] Elisha was from Hopkinton, Rhode Island, Samuel's hometown and had enlisted in Massachusetts to fight in the war. Elisha was related to Samuel via the Clark (also spelled Clarke) family of Hopkinton. The Clark family was a well-known and prominent family in the cities of Hopkinton, Westerly, and Mystic. Likewise, Samuel's mother's ancestors on both sides were related to the Clark family. Elisha and Samuel were in fact distant cousins, being fourth cousins five times removed from their common ancestor, Captain Lawrence Clarke of Rhode Island (1676-1754). These two men shared not only a bond of wartime camaraderie, but also as cousins and natives to the state of Rhode Island.

The second major person introduced in chapter four is Zerah E. Cottrell. Zerah E. Cottrell (1838 – 1922) was the Sergeant Major of Company B of the 23rd Iowa Infantry Regiment.[2] Zerah was originally from Hornell, New York, and lived in Xenia, Iowa shortly before the war.[102] Zerah enlisted as

---

[1] N. A. Strait, *Roster of All Regimental Surgeons and Assistant Surgeons in the Late War, With their Service, and Last-Known Post-Office Address*, (Washington: N. A. Strait, 1882), 101.

[2] *Roster and record of Iowa soldiers in the War of the Rebellion, Together with Historical Sketches of Volunteer Organizations, 1861-1866, Vol. III, 17th-31st Regiments-Infantry*, (Des Moines: Emory H. English, 1908), 699.

a Private in 1862 and was a wagoner (teamster) for the 23rd Iowa.[148] Zerah quickly rose through the enlisted ranks from Corporal to Sergeant, then from Sergeant to Sergeant Major of his company. Zerah was wounded on May 17, 1863 during the Battle of Black River Bridge and mustered out July 26, 1865, Harrisburg, Texas. Zerah died in 1922 at the age of 84 in Black Hawk County, Iowa, where he is buried. Strangely enough Zerah was married to Mary A. Palmiter (1841-1828), the daughter of Orsamus Palmiter (1809-1887), who was the brother of Samuel's fellow Albion friend Edwin S. Palmiter, who had enlisted with Samuel in 1862 and subsequently died in Vicksburg in 1863.

ENTRIES

Jan 1st, Sunday and New Years day . We fired a salute from the forts today of 100 guns in honor of the capture of Savanah Ga. By Gen. Sherman. Pleast.

Jan 8th, We recd Christmas dinner today from the good people of Wisconsin and were very thankful to them for the favor. Pleasant.

Jan 13th, We had Regt. inspection today in heavy marching order by the district inspection. Pleasant.

Jan 20th, I was detailed today with 75 men from our Regt. and the same number of cavalry to go down the river to look after some guerillas who had been firing on our boats as they passed up and down the river. We started down to Friers Point (12 miles) where we lay on the boat all night.

Jan 21st, We landed early this morning and our cavalry started out into the country to see what they could find and we lay in the town during the day. Our cavalry went out 22 miles with out meeting the enemy so they went into camp for the night and we lay at the town. It rained all day and night.

Jan 22nd, We lay at the town all day. Our cavalry came in the afternoon without any fight, caught 5 guerillas while they were out. We got on board the boat and started back to camp. We arrived at 10 oclock in the evening. It snowed a little during the night. Nothing of importance this month out. Cold.

Feb 10th, Our Regt sent out a scout of 50 men today with some cavalry They went down to Frier's Point on the boat and went down back of the town. Some of our cavalry had a fight with some guerillas. We had 1 man killed, 4 wounded and 3 captured. The expedition returned to camp the next day without any further excitement.   Pleasant and quite cold.

Feb 11ᵗʰ, Our Regt sent out another scouting party today of 50 men with some cavalry. They went up the St Francous river o boat and were gone 4 days. They had good success. They captured 21 prisoners with a loss on our side of 1 man captured.

Pleasant and cold.

Feb 15ᵗʰ, I recd $25.00 in a letter from G. F. Lawton today. Pleasant weather + cold.[3]

Feb 19ᵗʰ, Sunday. I was detailed with 50 men from our Regt to go down to Frier's Point with the Gen. as escort. We started out at noon on the steam boat Kate Hart . We run down in 1 hr and staid there until 4 oclock and then went back to camp where we arrived at 6 o'clock. Pleasant.

Feb 20 + 21ˢᵗ, In camp. Nothing transpired of importance. Rainy + cloudy.

Feb 23ʳᵈ, In camp. We got orders this morning to pack up and go on board the boat at nine o'clock A.M. We went on board the steamboat Fanny Ogden, and started out from Helena at noon We run down to Island No 63 where we tied up.[4]

Feb 24. We started on down the river early this morning and run down to the mouth of the White River where we tied up for 2 hrs and wooded up the boat again and then started on down the river again and run down to Lake Providence where we laid up for the night . It rained all day with a heavy thunder shower threw the night.

Feb 25ᵗʰ, We started on down the river early this morning. We run down about 26 miles and tied up again to wood up the boat. We stopped for 3 hrs and then started on down the river. We arrived at Vicksburg at 5 o'clock in the afternoon where we stopped of 5 hrs + coaled the boat. We started out for Vicksburg at 10 in the evening and run all night. Pleasant and very windy.

Feb 26 We passed Natchez at 6 this morning and the mouth of the Red River at noon and arrived at Morgansia at 3 in the afternoon where we stopped for 30 minutes. We then went on down the river and arrived at baton Rouge at 7 in the evening where we stopped for 20 minutes and then started on down the river and run all night. Pleasant all day and warm.

---

[3] The $25.00 sent to Samuel was from Giles Franklin Lawton, his friend and fellow Christian back in Albion, Wisconsin. This amounted to roughly $450 adjusted for inflation.

[4] Friars Point was a small settlement on the Mississippi River and a frequent stopping point for the 23ʳᵈ Wisconsin among other regiments to load wood and coal onto their steamboats.

Feb 27th, We arrived at New Orleans at six this morning and lay on the boat at the dock till noon. We then steamed up and run over cross the river to Algiers where we landed and went into camp out back of the town. It rained all day and night and we had a nasty time getting our tents up.

Feb 28. In camp. We had Regt inspection today in heavy marching order by the corps inspection and was also mustered in for 2 months pa. We got orders to get ready to march this afternoon but not being ready the order was counter marched and we remained in camp. It rained in the morning and pleasant in the after noon.

Mar 1st, In camp. We got orders this forenoon to pack up and get ready to march at 4 P M. We started out and marched down to the ferry where we crossed over to the city and marched out threw Canal Street and out to Lake End (a distance of 10 mile) and embarked on board the transport James Battle for Dauphin Island at Mobile Bay (at 9 oclock in the morning) We got the boat out of the canal into the lake and anchored till morning. Cloudy.

Mar 2nd, We got up stream early this morning and run down to the railroad landing to coal up the boat. We lay at the Dock till 3 in the after noon when we started out and run down cross Lake Pontoftrane and passed out at Port Pike at 7 in the evening into Miss . Sound. We passed Ship Island at 9 in the evening and anchored at Rocky Island at 2 in the morning on account of fog . Cloudy.[5]

Mar 3, Morning. Laying at anchor off Rocky Island. The fog cleared at nine + we started on and arrived at Dauphin Island at 3 in the afternoon We landed at Fort Ganes and went out back of the fort . and went into camp with the balance of our brigade which his composed of the 161st N.Y. , 29th Ill, 30th MO. , and our own . Regt which is called the 3rd brigade of the 1st Division of the 13rd Army corps. There is plenty of troops here and nobody knows what is going to be done. The distance from New Orleans to this place is 140 miles

Cloudy and a heavy thunder shower threatened.

Mar 4th, In camp. We set up our tents today and did some writing. ~~Pleasant~~ Cloudy and rainy.

Mar 5 Sunday In camp. I went to meeting today .   Pleasant.

---

[5] Here Samuel spells Lake Pontchartrain phonetically. Lake Pontchartrain is the largest estuary of the Mississippi River in Louisiana, noticeable from most maps.

Mar 6th, In camp at Dauphin Island . We remained in camp here till the 17th and drilled and fitted up for Spring campaign. ~~M~~

Mar 12 Our naval fleet went up toward the city of Mobile today and shelled the rebel land batteries all day with out much effect.

Mar 14 I was on detail today with 12 men to unload boats. It rained all day and night.

Mar 15 We had brigade inspection to day in heavy marching order today and also had a heavy thundershower. There was 1 man killed and 2 others badly injured by lightning.

Mar 17. We got orders this morning to get ready to move with 2 days cooked rations. The troops began moving today up toward the city of Mobile and we are to follow them as soon as we can get transportation across the bay. We broke camp this afternoon at 5 oclock and marched down to the landing (2 miles) and embarked on a steam boat. We started out from the wharf at 10 in the evening, crossed over and landed at a place 4 miles above Fort Morgan called Navey Coe and lay on the beach until morning.[6]

Mar 18 We started up the beach this morning at 8 oclock , marched 4 miles and went into camp. Gen. A J. Smith's forces landed on Cedar Point today to draw the enemies' attention that way while we went the other way We had a skirmish with the enemy.

Mar 19. We broke camp this morning and marched up the Peninsula 14 miles and went into camp for the night in heavy pine timber. Pleasant.

Mar 20 We broke camp this morning at 6 o'clock and marched 8 miles when we found that we were on the wrong road and had to march back 2 miles where we went into camp. I was on detail with 150 men to build corduroy road so to get our teams and artillery a long We built about ½ mile, there was a detail from all our division. Rainy weather[7]

Mar 21st, We remained in camp today and our division built corduroy toad all day by detail. We build about 2 miles. We had a heavy thunder storm and it rained nearly all day.

---

[6] The 23rd Wisconsin's landing was at Navy Cove, a cove just west of Fort Morgan, which had been captured by Union troops one year prior to the Battle of Spanish Fort and the Mobile Campaign.

[7] A corduroy road, also called a log road, is a road made from fallen logs and timber in muddy and low areas in order to improve the traction of dirt roads when wet. Corduroy roads have been used for much of human history. During the American Civil War corduroy roads came to prominence throughout a variety of battles and campaigns and were used by both sides.

Mar 22. We broke camp at 6 oclock this morning and marched 6 miles during the day. The roads were very bad and we were on detail most of the day building road and getting our teams out of the mud. We camped for the night in heavy pine woods.   Pleasant.

Mar 23 We broke camp this morning at 6 and went to building roads, and geting our teams along . We built about 2 miles when we stopped at noon, drew rations, got out dinner and went to work on the road, we built and got along 2 miles more and went into camp for the night. We got supper and went to work, on the road again and worked till 11 o'clock at night. Pleasant.

Mar 24 We broke camp this morning at 6 oclock and marched to Fish River and 2 miles beyond, marching 10 miles and went into camp in the woods. The roads were quite good today, having got on high ground. Our division has built 10 miles of corduroy road coming thru to this place. We found A. J. Smiths forces in camp here, they having come up to this place on transports. Our advance had a skirmish with some rebs yesterday near this place with out much loss on either side. Pleasant.

Mar 25th, In camp at Fish River. We got orders to get ready to march at noon with 5 days rations in haversack. A. J. Smith's corps commenced moving out at noon. Our Brigade moved out at 3 oclock. Our Brigade guarded the train today, we marched 7 miles and went into camp for the night in heavy pine woods. Pleasant all day.

Mar 26 Sunday. We broke camp at 6 oclock this morning and moved out about 6 oclock this morning and moved out about 6 miles when our advance met the enemy and commenced skirmishing with them. We skirmished about 2 miles when we came up to the main force from their fortifications skirmishing stopped for the night and our enemy went into camp for the night. Our Regt was sent out to the for on picket during the night. Pleasant.

Mar 27. Our army commenced moving up to the fort this morning at 7 oclock and was attacked by the enemy in force but our boys soon routed them and drove them into their fortifications Our Regt. was relieved from picket and skirmishing at at noon and we took our place in line of battle of our brigade. We marched up to with in ½ mile of the Rebel forts and threw up a line of rifle pits for our protection. The rebels shelled our lines all day and our artillery replied in good earnest. We had several men killed + wounded during the day. Made 6 miles today. Rained afternoon.

Mar 28 We lay in line of battle all night last night The rebels were busy all night strengthening this works and getting reinforcements from the city of

Mobile The reb opened on us this morning at 6 o'clock with this artillery and skirmishing. We soon were ready and replied to them with the same. There was considerable artillery and picket firing all day with considerable loss on both sides. We abandoned our lines several hundred yds today. Our Regt was sent out the fort at dark to hold the picket line during the night. We relieved the 23rd Iowa Regt. It was very quiet along the lines during the night.

Mar 29. Our Regt was relived from picket this morning at day light by the 20th Wis. Regt. There was no excitement during the night. We marched back to the rear of our lines this morning to get our breakfast and some sleep but the rebels shelled our position so bad that we had to move 1 miles to the right to get under the cover of a hill. Our loss was very heavy today in killed and wounded. We had 4 men badly wounded in our Regt today. The enemy kept up a continual fire all day One of our Monitors was sunk by a torpedo yesterday and another is reported sunk today. Our fleet is clearing the channel of torpedos so as to get up in shelling distance of the rebel forts. They threw a few shells today at the fort

Mar 30 The rebels came out of their works last night at 11 oclock and made a dash on our outer pickets and men who were throwing up rifles pits in front of the works and above them back to our main lines when or boys gave them a few volleys that sent them back to their works at a double quick and they did not trouble us again during the night. Our loss was very slight The enemies must have been heavy. The rebels shelled us all day furiously and our front line was kept very busy keeping the rebels down behind their works. Our army has been busy all day getting our morters and heavy guns into position as it is evidint we came to shell them out. Fort Spanish is situated on the Kenesaw River near its mouth and is well guarded by outer works. It is sand the garrison numbers from 7 to 8000 men. It is 12 miles from Mobile by water and 30 by land. Our division got orders this afternoon to move back to the rear 2 miles, draw 4 days rations and go as an scout with a supply train to Gen. Steels army that is said to be 13 miles from here towards Mobile and our of rations. We were relieved from the front at dark by some of Smiths forces and moved back two miles to the rear and went into camp for the night and drew 5 days rations.   Pleasant.

Mar 31st, Our division broke camp this morning at 6 o'clock and started out with a large supply train in the direction of Blakly. We marched out 7 miles and went into camp at noon. We went to work and threw up a line of rifle for our protection in case a sudden attack from the enemy. There was heavy firing in the direction of Fort Spanish and it is supposed they had heavy fighting all day. The country that we have marched threw so far is heavy pine woods and sandy soil. Pleasant.[8]

Apr 1ˢᵗ, In camp 7 miles from Blakely. We remained in camp all day waiting to hear from Steels army. There was very heavy firing at Fort Spanish all day. A Brigade of our cavalry came up from Navey Coe today. They went out in the country 19 miles this after noon but they did not see any enemy.

Apr 2ⁿᵈ, Sunday. We had company inspection this morning and also attended Church. Gen Steels advanced forces came in to us at noon today and reported his army allright at Blakley. Our supply train was unloaded and the train was sent back to the landing this afternoon for more rations. There had been heavy firing at Fort Spanish all day and in the direction of Mobile. Gen Steel's train came in this after noon for rations. His forces have been fighting at Blakely today. There is heavy firing in that direction this evening The 14 brigade of our division went out to Blakley this evening to reinforce Steel. It is reported that A. J. Smith sunk one of the rebel transports at Fort Spanish today. There has been heavy firing in that direction. Pleasant.

Apr 3. Our first brigade went out to Blakley last evening and our brigade was routed at last night at mid night to follow them. We started at midnight and march 5 miles stacked arms ad lay down to sleep. We lay one hr. then got up and marched up to the rear of Steels army, stacked arms and got our breakfast and waited for orders We got orders at noon to march out to the left of steels line of battle and from on his left ~~of Steels~~ where we remained until 3 in the afternoon. Several rebel shells went over us today but nobody was hurt in our Regt.. At 2 oclock the 2ⁿᵈ division of Smiths corps came up and took our place and we marched back 2 miles to the rear drew rations and went into camp for the night. Steels fore captured 400 prisoners + 4 pieces of artillery coming up to this place and two rebel Gens. There was heavy firing at Fort Spanish all day and also at this place. Pleasant

Apr 4ᵗʰ, Every thing remained quiet last night with the exception of a few heavy guns on both sides We remained in the rear today and rested. Every-thing was very quiet until 5 o'clock P.M. when our heavy artillery and morters with our gun boats opened on Fort Spanish and kept up a furious bom-bardment for 3 hrs. The rebels di not reply much as our boys did not give them a chance We had orders to be on the lookout this evening as there was a rebel force trying to get into our rear but I guess they did not come

---

[8] "Blakly" is the phonetic spelling of the city of Blakeley, Alabama, in Baldwin County, Alabama, which is now a ghost town. Blakeley is also the location of the Battle of Fort Blakeley in 1865.

as we did not see them. There was considerable firing on both sides during the night. Pleasant

Apr 6th, We were called our last night at 10 oclock to go to the front as it was ascertained that the rebels were trying to break threw our lines. They made 3 charges on our front lines during the night but were easily repulsed by our boys. We marched back to our camp this morning at sunrise, not being needed in the front (making 8 miles march). We remained in camp in the rear all day and washed up our clothes There was a salute of 100 guns fired by the navy this afternoon in honor of the capture of Salma and the defeat of Forest by Gen Thomas. There was considerable firing all along the liens today. Pleasant.[9]

Apr 7th, There was considerable firing from both today. Our Regt. was detailed this after noon to go to the front on fatigue during the night to build works for our heavy artillery. We started out at dark and went to the front on the right of our line (about 4 miles from our camp). The rebels found out where we were to work and shelled us severely at spells all night but we were very fortunate, we only had one man killed + 2 wounded. We worked all night and went back to our cam at day light. Pleasant.

Apr 8th, We lay in camp all day There was heavy firing all along the lines. The rebels used their artillery very freely today. Our artillery broke loose down to Fort Spanish this afternoon at 5 o'clock and kept up the most terrific bombardment I ever heard for 3 hrs. and I guess there were soreheads in front of Fort Spanish. There was considerable firing at Fort Spanish during the night + also at this place . Cloudy + cold.

Apr 9th, Sunday in camp. Everything is quite still this morning with the exception of a little artillery firing on both sides. The Rebels surrendered at Fort Spanish this morning and our men took prisoners at 9 A. M. We captured 687 prisoners and 11 pieces of artillery The rebel loss in killed + wounded was 620 and our loss was 250 killed + wounded. At about noon today the Gen ascertained that the rebels had commenced to evacuate Fort Blakly so he ordered every thing to be ready to make a charge at 2 oclock. We were delayed a little so the artillery opened fire on the fort about 4 o'clock for 15 minutes when our boys made the charge on the fort in good style and captured the fort in about 15 minutes with 400 prisoners and 50 pieces of artillery. Our loss in killed and wounded was about 200 and the rebel loss

---

[9] The capture of Selma, Alabama, on April 2, 1865, and the subsequent surrender of Confederate Cavalry General Nathan Bedford Forest was actually cedited to Union Major General James H. Wilson (1837-1925), and not General Thomas.

about the same. So ended the siege of the two strong holds and our way was clear to the rear of Mobile. After the capture of the fort our Regt. ~~of~~ went back into our old camp for the night.   Pleasant .

Apr 10. We had orders this morning to be ready to march this morning at sunrise. We broke camp at 6 o'clock and marched miles out up the Kenessau River and went into camp for the night. We cleaned up our clothes

Apr 11th, In camp. I went down to the rebel works to take a look at them. Our guns boats. Our gun boats had been shelling some rebel batteries up toward the city all day. It is reported that the city offered to surrender on condition that we would protect personal property with out any fighting but Gen Camby says no conditions .

Apr 12th, We got orders last evening to pack up ready to march as it was ascertained that the rebels were evacuating Mobile. We got ready and marched at 7 in the evening and marched all night and arrived at the landing below Fort Spanish at 5 this morning. The ~~14~~th + 3rd divisions of our corps embarked on transports and started across the bay at 9 oclock this morning (about 8 miles across the bay) We landed 6 miles below Mobile when we disembarked which took us till 4 oclock as the piers were all torn away. As soon as the troops were all landed we started up toward the city and went into camp at 8 in the evening. We had just got our coffee for supper when the order came for us to move up into the city and took possession and went into camp at 10 oclock at night. The rebels destroyed all they could not take away with them. Pleasant

Apr 13. In camp near the rebel fortifications in the rear of the city. I went around the line of rebel fortifications this forenoon to take a look at them. They are the strongest lines of works I have ever seen. The rebels spiked all the guns and burnt all carriages. There was 460 guns in the fortifications around the city and a large lot of ammunition all fell into our hands. A large lot of the rebels stand in around the city and would not go away with the rebel army. Our boys are busy picking them up and taking them to headquarters where they will probably take the oath and be allowed to remain in the city. A good union man took several of us boys into his house and gave us a good dinner. We moved our camp this after noon out to the main line of works in the ~~rear~~ city and went into fix camp and fixed up our quarters. I went out to a house today and got all the strawberries I could eat . Pleasant

Apr 14, In camp I went down this for noon to take a look at the city. It is quite a nice city Our fleet came up to the city yesterday bringing our trains, baggage and rations It is reported today that our cavalry that went out in the rear of the ~~rebel army~~ city ~~rebel~~ captured all of the rebel army that left here and all of the gun boats and transports + artillery They are bringing them down the river today Our losses during the operations around Mobile have been about 1000 men killed and wounded and missing. and the rebels loss about 6000 and 500 pieces of artillery. We had one Monitor sunk and one disabled and one tin clad sunk during the capture of Mobile. Deserters are still coming in in large numbers. I was detailed on picket this afternoon. Pleasant.

Apr 15. In camp. I came in from picket. Got some mail today Pleasant.

Apr 16. Sunday. In camp. We had Regt inspection in heavy marching order this forenoon. I got some letters today . Pleasant

Apr 17-18-19. In camp. Nothing of importance transpired except the firing of 200 guns over the new of Gen. Lees surrender to Gen Grant at Apt Pleasant

Apr 21st In camp. Nothing transpired of importance today.

22nd In camp. We received the sad news of the assassination of Pres. Lincoln. Our flag was lowered to half mast and we fired minute guns all day over the sad news.

Apr 23rd, Sunday in camp. We had Regt inspection in heavy marching order by the division inspector. I was detailed on picket . Pleasant

Apr 24 In camp. Came in from picket this morning. All quiet. Got some mail + wrote some letters Pleasant.

Apr 25 In camp. Went down to the city. One of our transports run on to a torpedo today coming up the bay and was blown up. There was 7 persons killed and several wounded. Pleasant

Apr 26 In camp. Gen. Dick Taylor, Adj. Gen. came in to the city today under flag of truce to see on what terms he could surrender his army to Gen Canby but Canby was in New Orleans so he had to go there the see him . It is supposed hat he will surrender in a few days   Pleasant.

Apr 27th, In camp. Nothing transpired today I was on guard.

Apr 28th, In camp. Came off guard. Our division commander was assigned to the command of the city today (cloudy)

Apr 29. In camp. I went blackberrying. Had first rate luck.

Apr 30 Sunday in camp. Had inspection of arms this morning The Regt was mustered for 2 months pay. Also signed roll for 10 months pay. Pleasant

May 1 In camp. Came in from picket this morning Another of our transports was blow up today by running on to a torpedo. No lives were lost .

May 2 In camp. I was detailed on guard today. Nothing of amount transpired today .

May 3 In camp. Came off guard this morning I received 10 mo. pay today and I settled up all my accounts. Pleasant + hot.

May 4 In camp. I went blackberrying. Pleasant + warm

May 5 In camp. Dick Taylor has surrendered his army today and the war is now over. I was on detail today getting ready to move our camp tomorrow.

May 6. We moved camp today about 1 mile over on to Government St. and put up tents etc. Had a visit from Elisher Pollark today. Cloudy + rainy thru the night.

May 7 Sunday. In camp are finished fixing up our camp and changed up the sheets today.

May 8 In camp I was detailed on fatiege today to take the heavy guns off the rebel works. We presented Major Green with a horse today for which we paid $500.00 . Pleasant .

May 9. In camp. I went down in the city today and had a good visit with E. P. Clark of the 6th Mass. I expressed home one hundred dollars today. Pleasant + warm.[10]

May 10. In camp. I was detailed on picket today. Nothing of importance transpired today. Showers.

May 11 In camp. Came in from picket, got some mail and did some writing

May 12 In camp. I was detailed on guard today, had dress parade this afternoon. Pleasant.

May 13. In camp. Went down to the 23rd Iowa Regt to see Zera Cottrell. I went out blackberrying this afternoon. Pleasant.[11]

May 14. Sunday in camp. Had inspection of arms in the morning + dress parade in the afternoon

May 15 + 16 In camp. Nothing transpired of any importance. Pleasant + very warm.

---

[10] E.P. Clark refers to the aforementioned Elisha Peckham Clark, one of Samuel's cousins from Rhode Island.

[11] Here the name "Zera" is spelled phonetically. Samuel is referring to Zerah E. Cottrell, a Sergeant in the 23rd Iowa.

May 17th, In camp. We got news of the capture of Jeff Davis and his cabinet by Gen. Wilson's cavalry.

May 18-19-20 In camp. On picket .

21st, Sunday. In camp. We had inspection of arms this morning. I did some writing in the afternoon. Pleasant.

May 22-23. In camp. I was detailed on guard today had a visit with E. P. Clark, Surgeon of the 6th Mass Regt.   Pleasant .

May 24th, In camp. Came off guard and did some washing today.   Pleasant .

May 25. In camp. I was detailed on picket this morning. There was an explosion occurred down town this afternoon at 2 oclock. The Ordnance Building was accidentaly blown up causing the loss of from 3 to 4 hundred lives and a large number of wounded It contained about 30 tons of gunpowder and a large quantity of shells and ammunition. The explosion destroyed five blocks of large buildings mostly brick and damaged the city for a mile around. The soldiers were turned out to manage fire engine and the destruction by fire was soon stopped. Pleasant + very hot.

May 26. In camp. We had 18 men wounded in our Regt yesterday by the explosion and a large number of our soldiers were killed and wounded. Our Regt. got orders to be ready to go to New Orleans as soon as we can get transportation. Pleasant.[12]

May 27. In camp. The explosion yesterday was the greatest ever in this country. It caused the destruction of 6 blocks of large brick buildings. It jarred doors and windows for a half mile around. There was about 2000 killed and wounded and a very large destruction of property. The arsenal contained 800 tons of shot and shells. There was several buildings burned.

May 28 Sunday. In camp. Was detailed on on guard today. Nothing of importance transpired. Pleasant.

May 29 In camp. Came off guard. Went down town and went to the theater in the evening. The 1st + 2nd Brigades of our division went to New Orleans today. Pleasant

May 30 In camp. Went out in the country visiting I had a good time and got all the black berries I wanted. Pleasant + hot.

---

[12] The explosion above that Samuel describes was the Mobile Magazine Explosion, which destroyed much of North Mobile, sank two ships, and killed upwards of 200 people.

May 31. In camp.   Pleasant .

June 1 In camp. Went down to visit the 31st Mass. Regt today   Pleasant.

June 2. In camp. Was detached on guard today. Col. Guppey came back to our Regt today.

June 3. In camp.   The troops in and around the city were reviewed this morning at 8 oclock by Maj. Gen. Granger and Clnel Justin Chase. The review was very satisfactory to all who saw it. There was a very large turn out. Our Regt was complimented very highly . Hot + dusty

June 4. In camp. Sunday. We had inspection of arms this morning. I did some writing. Had dress parade this evening. Pleas

June 5 6. 7 + 8th, In camp . Pleasant + hot .

June 9 + 10th, In camp .

June 11th, Sunday in camp. Was detailed on guard. I went to the Willis + Thompson's Minstrells last evening . Pleasant and hot.[13]

June 12th, In camp. Came off guard, got some mail. Heard from my money sent home the 10th of May. Thunder storm today.

June 13 In camp. Had dress parade this after noon. I went to the academy of music this evening. Showery today

June 14 + 15th. In camp. Pleasant.

June 16. In camp. Gov. Lewis of Wisc. Visited our Regt. today. We had dress parade after which the Gov. made a short speech to us and complimented our Regt. Pleasant + hot.

June 17 In camp. Pleasant.

June 18. Sunday. In camp. We had inspection of arms this morning and dress parade in the evening.   Pleasant.

June 19. In camp. Had dress parade in the evening Pleasant.

June 20th, In camp. We had dress parade today for the benefit of Gen. Andrews. He was well pleased. Pleasant .[14]

---

[13] Willis & Thompson's Minstrels was a Minstrel Band which had been founded in Mobile, Alabama, on May 27, 1865. According to the book *Burnt Cork and Tambourines: A Source Book of Negro Ministrelsy*, the bandmembers were as follows; Oscar Willis, Willis Armstrong, E. K. Collins, James Snyder, C. G. Spinola, Master Bobby, J. W. Thompson, James Livingston, John Stuart, James Weaver, Ed Magruder, J. T. Murphy, as well as Frank and Emil Sigel.

[14] Christopher Colombus Andrews was a Union General who commanded the 2nd Division of the Union Army's XIII Corps. Andrews was originally the Captain of Company I of the 3rd Minnesota Infantry Regiment and later commanded the 3rd Minnesota before his appointment to Brigadier General.

June 21. In camp. Came off guard this morning. Gen Andrews inspected our camp today. Had dress parade in the evening. Pleasant.

June 22 23 + 24 In camp. Drilled and had dress parade.   Pleasant.

June 25. Sunday   In camp. Had inspection in the morning. Had heavy thunder showers in the afternoon + evening.

June 26, 27, 28 + 29th   In camp. Had drill and dress parade.   Pleasant.

June 30th**,** In camp. The Regt. was mustered in for 2 mo. pay. We got orders today to make out our muster rolls to be mustered of the service. Rained afternoon

July 1st**,** In camp. We turned over our today to the 35th Wisconsin Regt. Got some mail and wrote some letters. Rainy

July 2+3rd**,** In camp.

July 4th**,** In camp. Our Regt. was mustered out of the U.S. Service at 6 oclock this morning by Leiut. J. L. Baker. There was not much doing today. Pleasant.[15]

July 5. In camp. We got orders this morning to get ready to go on board the steam boat Warior for New Orleans. Went on board at 6 oclock and started out from Mobile at 7 oclock in the morning. We ran all day and night, 160 miles. Pleasant.

July 6. We arrived at Lake End this this morning at 4 oclock. We then embarked on board the cars and run over to the city of New Orleans where we arrived at 7 o'clock. We lay at the Depot here until noon when we went on bard the steamer Capt. Robinson for Cairo. We started out at noon and run up to the coal yd to take on coal which took us until 5 o'clock. We got ready and started up the river at the foot of Canal St. to get rations for the boat. We got ready and started up the river at 20 oclock at night and run all night. Pleasant.

July 7th**,** We passed Donaldsonville this forenoon at 10 and arrived at Baton Rouge at 4. in the afternoon when we stopped for 15 min and then started up the river, Passed Port Hudson at 8. Morgansia at 12. Red River at 4 oclock at night. Pleasant.

July 8. We stopped 1 hr this morning on account of the fog 25 miles above the mouth of the Red River, After the fog cleared up and we started on up the river. We arrived at Natchez at 2 oclock in the afternoon where we stopped for 30 min. and then went up the river and run all night . Cloudy and rainy.

---

[15] James L. Baker (1835 – 1901) was originally Company D's First Sergeant. Baker later became the 1st Lieutenant of Company D on February 6, 1863. Baker moved to Kinmundy, Illinois, after the war and was active in the Grand Army of the Republic (GAR) as a member of the local GAR Post 255 (Hicks) in Kinmundy. Baker died on June 11, 1901, in Kinmundy where he is buried (see image in the middle of book).

July 9th, We arrived at Vicksburg at 8 this morning and stopped for 4 hrs to draw rations and coal up the boat. We left Vicksburg at noon and run all day + all night . Pleasant + hot.

July 10. We passed Greenville this morning at 8 oclock in the evening, passed Napolian at 8 in the evening. Arrived at White river Landing at 10 and stopped 1 hr, went on up the river and run all night. Rained afternoon and evening.

July 11 We stopped at Island No 66 this morning at 8 for 30 min. We arrived at Helena at 2 this afternoon, stopped 2 hrs to take on coal. Left Helena at 4 oclock and run all night. Pleasant.

July 12. We arrived at Memphis this morning at 7. and stopped 1 hr. to draw rations for the boat. Started on up the river Passed Fort Pillow at 7 oclock in the evening and run all night. Pleasant + hot .

July 13. We passed New Madden at noon today Island No 10 at 2 oclock, Heckman at 6 and Calumet at 9 in the evening and arrived at Cairo at ~~at~~ 12 oclock at night We lay on the boat until morning. Pleasant

July 14. We got off the boat at 6 this morning and went on board. cattle cars at 8 o'clock We left Cairo at 8:30 in the forenoon , Arrived at Barberville at 3 oclock 65 miles. Waited for train 30 min. started on and arrived at Centralia at 6 oclock and stopped 1 hr. started on at 7 oclock. Passed Vandalia at 9 oclock and run all night . Pleasant fore noon and rained afternoon + night

July 15. We arrived at Minnonia at six this morning . Being 132 miles and stopped for the night . We started on in the night Lassalla at 9 o'clock. 50 miles from Minnonia We stopped 30 min. and started on for Beloit. Passed Amber at 12, Dixon at 2 and arrived at Freeport at 4 and stopped 1 hr. Arrived at Beloit at 7 in the evening where we remained over night and the people gave us a good supper. Distance from Freeport 55 miles

July 16. We left Beloit at 7 oclock this morning and arrived at Madison at 9 this forenoon where the people had a good breakfast waiting for us. After breakfast we marched up thru the city and was received in the Capital Park by Officials of Madison after which we marched up to Camp Randall where we left our camp equipage and was allowed to go where we pleased. The 6th + 7th Regt came in this evening and were received in the park. Pleasant.

July 15 The Regt was got together today at 10 oclock to turn the Government property in its procession and to sign the pay role and was then allowed to go home We went home in the afternoon and staid until the 25th with the Regt got together at Madison and was paid off and discharged from the U.S. Service and we all went home as citizens. Pleasant and every thing all night.

* * *

Journal kept from 1862 to 1865 by Corporal Samuel Burdick
Company D. 23rd Regt Wis Inft. Vol.

Presented to Dr Herbert R Bird
drummer boy of Co D. 23rd Regt Wis Vol.
May 23rd 1910

By Corpral Sam Burdick of Co D 23rd Regt Wis Vol. Inft.
Comrad Samuel Burdick.
Albion Dane Co Wis.

The distance traveled by the Regt
Marched by land – 1630 mils
By Rail Road – 1435 "
By Water on Boats + vessls – 7430 "
                    Total 10495 "

Distance from Madison Wis to Cairo Ill. From Cairo Ill to the Miss River,
to Columbus 350 mils
" – Memphis – 220 "
" – Hellena – 450 "
" – White River – 620 "
" – Greenville – 775 "
" – Lake providence – 805 "
" – Vicksburg – 860 "
" – Natchez – 980 "
" – Red River – 1040 "
" – Morganzia – 10470 "
" – Port Hudson – 1100 "
" – Donaldsonville – 1170 "
" – New Orleans – 1250 "
" – Fort Jackson to Philips – 1310 "
" – Mouth of the Miss River – 1350 "
Distance from New Orleans to Mattagarda Bay Tex, 600 miles
New Orleans to Mobile 140 "

The Albion Seventh Day Baptist Church, which Samuel and Albert Burdick repaired in the postwar era. *Courtesy of the Albion Historical Society.*

The Postbellum

Samuel returned to Albion in late July of 1865 after the 23rd Wisconsin Infantry Regiment was fully discharged from service and their equipment was returned to the regiment's storehouses at Camp Randall in Madison according to his July 15, 1865, entry. One year after the war Samuel married his wife Lucy Ann Saunders (1845-1912) in Albion on July 28, 1866.[1] Samuel and Lucy had three children altogether, two of which died in their infancy. Samuel's daughter May (Mae) Juliet Burdick (1870-1928) was his only child who survived infancy and would go on to marry Mahlon James Bolser (1863-1948) of the nearby town of Little Prairie, Wisconsin. Samuel continued his faith as the Seventh Day Baptist Christian, the same religion practiced by his family back in Rhode Island. As much of the populace of Albion was of "founding stock" or "Old Yankees" from Rhode Island and New York, much of the populace of the city were still ardent and practicing Christians from both the Albion SDB and the Albion Prairie Primitive Methodist churches.

After the death of his parents Samuel Sr. and Mercy in 1866, Samuel's younger brother Albert Clarke Burdick (1840-1919) followed in his brothers' footsteps and moved westward to Wisconsin after serving as an enlisted man and later a commissioned officer of the 5th Connecticut Volunteer Infantry Regiment and a short-lived business in New York City.[2] Samuel and Albert lived in the same home for quite some time, the two of them working as the town carpenters who built much of the city as it grew in the 1870's and 1880's, the brothers were also prominent local members of the Albion Seventh Day Baptist Church which they refurbished in the 1870's.[3] The Burdick house was

---

[1] *The Sabbath Recorder*, Volume 72, No 8, Feb. 19, 1912, 252.

[2] *The Wisconsin Tobacco Reporter*. (Edgerton, Wis.), 14 March 1919, Image 5. *Chronicling America*.

[3] Park, *Madison, Dane County and Surrounding Towns*, 290.

originally built around 1866 and still stands today in Albion, being situated just behind the remains of the Albion Academy at 584 Academy Drive. An inscription of Samuel's name can still be seen in the house's carriage garage. Samuel is listed in Albion's 1870 United States Census, as well as in his pension records as being employed as a house carpenter. Samuel's original pension request indicates that he ran his own carpentry and architectural business in Albion until the onset of advanced age, deafness, and poor eyesight which would ultimately hasten his early retirement from the carpentry trade.

Samuel's business card from his pension record

The Burdick house in Albion. *Courtesy of Daniel Scharfenberg.*

Both Samuel and his brother Albert were members of the local Edgerton post of the Grand Army of the Republic (GAR), a political, fraternal, and charity organization for Union veterans of the American Civil War. In its initial years when the GAR was first organized in 1866 in Illinois, it had very poor attendance nationwide as many veterans organizations were scattered and not well organized.[4] With the help of Union General Lucius Fairchild however, Wisconsin established its first GAR post in Madison, making Wisconsin the second overall state to establish a GAR department in the continental United States.[5] The GAR exerted political pressure on local, state, and national governments to further veteran's legislation and benefits, as well as demonstrate patriotism and charity to the public.[6] In 1878, the Wisconsin Department of the GAR had 31,016 members in total.[7] By 1890 Wisconsin GAR membership had risen to a booming 409,489 members statewide.[8] Wives and daughters of veterans also supported the GAR by creating the Women's Relief Corps (WRC) in Wisconsin in 1883.[9] Likewise, the Sons of Veterans (SOV or SV) served a similar purpose to the WRC for sons and grandsons of Union veterans as both a fraternal organization and paying tribute to the memory of Union veterans.[10]

The city of Edgerton in Dane and Rock counties organized the local GAR Post 137 "H. S. Swift", named after 1st Lieutenant Henry Spencer Swift Jr. of Company E, 33rd Wisconsin Volunteer Infantry Regiment. Lieutenant Swift had died on April 19, 1864, while fighting with the 33rd Wisconsin during an action at Perry's Ferry near Coldwater, Mississippi.[11] Edgerton veterans applied for a post charter and conducted its first official muster on February 11, 1884, with a charter of thirty-three total members.[12] Like many posts, the Edgerton GAR post was heavily active in the local community,

---

[4] Frank L. Kent, *Wisconsin in the Civil War: The Home Front and the Battle Front, 1861-1865,* adison: The State Historical Society of Wisconsin, 1997), 126.

[5] Kent, *Wisconsin in the Civil War*, 126.

[6] Ibid, 126.

[7] Hosea W. Rood. "The Grand Army of the Republic and the Wisconsin Department." *The Wisconsin Magazine of History* 6, no. 3 (1923): 287. http://www.jstor.org/stable/4630430.

[8] Ibid, 287.

[9] Kent, *Wisconsin in the Civil War*, 127.

[10] Ibid.

[11] Jeremiah M. Rusk and Chandler P. Chapman, *Roster of Wisconsin Volunteers, War of the Rebellion, 1861-1865. Volume II.*, (Madison: Democrat Printing Company, 1886), 511.

[12] Thomas J. McCrory, **Grand Army of the Republic: Department of Wisconsin**, (Black Earth: Trails Books, 2005), 207.

hosting many luncheons, parades, remembrance events, decoration days, and meetings for local veterans and their families. Royal Hall was the primary meeting location for the Edgerton H. S. Swift post of the Grand Army of the Republic. Royal Hall was a theater hall and meeting location for much of the city of Edgerton and surrounding communities around the turn of the 19th century before it was purchased by the Edgerton Order of Odd Fellows and Freemasons for their meeting center. The Henry S. Swift Post 137 lasted until 1936, the post's final member being John Sherman (1841-1932) of Fulton, Wisconsin who served in Company F of the 50th Wisconsin Infantry Regiment from 1865-1866.[13]

Besides his activity with the GAR, Samuel also met with many of his veteran comrades on a regular basis as a member of the Twenty-Third Reunion Association. Samuel was present at most of the 23rd Wisconsin Infantry reunions which took place in 1886, 1889, 1898, 1907 and 1912, although Samuel was not present at the 1889, 1898, and 1907 reunions, the organization still had Samuel listed as a survivor of the regiment in all reunion pamphlets.[14] Samuel was present at the dedication of the Camp Randall Memorial Arch commissioned by artist Lew F. Porter in 1912 and was one of the nine total survivors of Company D of the 23rd Wisconsin that attended the monument's dedication day.[15]

Veterans, their families, and onlookers gather at the dedication of the Camp Randall Memorial Arch on June 19, 1912. Samuel is in this photograph among the few remaining Wisconsin veterans of Company D of the 23rd Wisconsin. *Wisconsin Historical Society. Image 11270.*

---

[13] McCrory, *Grand Army of the Republic*, 207.

[14] Reunion of the Twenty-Third, 1907, 1919, *Wisconsin Historical Society.*

[15] Reunion of the Twenty-Third, 1912, *Wisconsin Historical Society.*

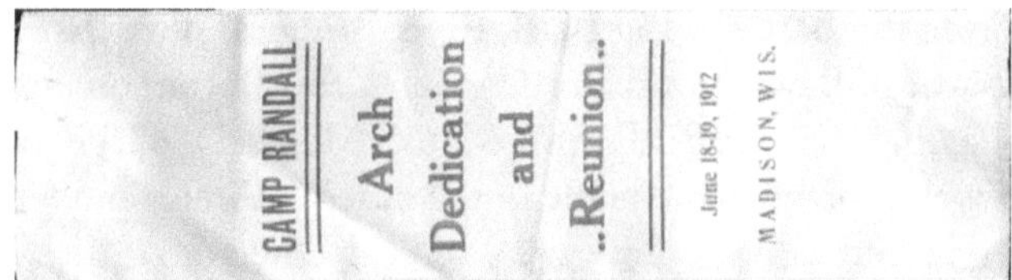

An example of one of the Camp Randall arch dedication and reunion silk ribbons given to attendees of the dedication in June of 1912. *Wisconsin Veterans Museum (WVM ID: V2002.1.32).*

A group portrait of members of the Dane County Veterans Association, a contemporary veterans organization of the Grand Army of the Republic open to veterans of the Civil, the Spanish American War, and soldier's wives. Members are wearing ribbons, and some are wearing badges. Amongst the group is a visible drum corps and two American flags. The bass drum appears to read: "Lucius Fairchild G. A. R. Post No. 11". This image was taken at Camp Randall. *Wisconsin Veterans Museum.*

One of the most important events pertinent to this book occurred on May 23, 1910 when Samuel gave his dairy to a fellow comrade who fought in his unit, Dr. Herbert Roderick Bird, who was previously mentioned at the end of the diary. Herbert Roderick Bird (1849-1923) was the drummer boy of Company D and had originally enlisted with the 23rd Wisconsin Regiment at the young age of thirteen in August of 1862.[16] Herbert was wounded on May

---

[16] Jeremiah M. Rusk and Chandler P. Chapman, *Roster of Wisconsin Volunteers, War of the Rebellion, 1861-1865. Volume II.*, (Madison: Democrat Printing Company, 1886), 240.

17, 1863 during the Battle of Big Black River Bridge and was later captured during the Battle of Bayou Bourbeux (Carrion Crow Bayou) on November 3, 1863.[104] Following the war, Bird went to Rush Medical College in Chicago, Illinois and would practice medicine in the cities of Arena and Madison in Wisconsin. Herbert later became the Medical Director of the University of Wisconsin-Madison and was heavily active in the Grand Army of the Republic, later obtaining the rank of Post Commander. Herbert was later the acting secretary of the 23rd Wisconsin regiment's survivor organization after the death of the organization's commander Lieutenant Colonel William Freeman Vilas. Herbert died in Arena, Wisconsin on June 25, 1923 at the age of seventy-four. He was preceded in death by his wife Gardia Ann Meigs and survived by his two sons Herbert R. Bird Jr. of Arena and Reverend William G. Bird of Illinois. According to the Wisconsin Veterans Museum's outreach and reference archivist Russell P. Horton, Samuel's diary was likely donated by Herbert Bird or his descendants after the original Grand Army of the Republic Memorial Hall burned down in 1904 as part of the broader 1904 Wisconsin state capitol fire. Immediately following the fire there was a call for Civil War veterans to donate letters, diaries, images, etc. to the newly created museum which is now the Wisconsin Veterans Museum.

(l) Sketch of Herbert R. Bird c.1862 as Company D's drummer boy. Courtesy of the Wisconsin Veterans Museum; (r) Dr. Herbert R. Bird in the 1910s– 1920s at a much older age. *Courtesy of the Wisconsin Veterans Museum.*

Pension, Health, and Death

The postwar era left hundreds of thousands of Union and Confederate veterans disabled both directly and indirectly from injuries sustained during the war. While veterans grew older, many struggled to work jobs, trades, teach, or farm due to injuries they sustained years prior during wartime service. Indeed, many could not work at all. The first major political action to allow Union soldiers a pension was the Dependent and Disability Pension Act of 1890, also called the Sherwood Act. This act allowed all veterans of the Mexican-American War, as well as Union veterans of the American Civil War to be granted a pension regardless of their disability once they reached the age of sixty-two.

Samuel's pension card. *Courtesy of the National Archives and Records Administration.*

Samuel first filed for a veteran's pension on November 14, 1888. Like many veterans Samuel was physically disabled from the war as soon as he returned home to Wisconsin. To afford medical care costs, as well as acquire benefits for his military service during the war, Samuel applied for a disability pension due to a combination of ailments directly caused by his time in the service. Samuel's pension record indicates there was "total deafness in right ear and severe deafness of left ear" as well as having "disease of eyes". That is to say that Samuel was almost completely deaf and nearly blind. The basis of Samuel's pension states that his deafness came from a "concussion of artillery at Vicksburg Miss May 63 and Jackson Miss July 63", and that his partial/ full blindness started in Mobile, Alabama, when he was discharged from the service in July of 1865. "Disease of eyes" could indicate that Samuel was fully or partially blind either due to glaucoma or cataracts, more than likely the latter. The pension descriptions match Samuel's journal kept during the war during May to July of 1863 and is corroborated by his participation in the main assault made on the breastworks and forts at Vicksburg when the 23rd Wisconsin was exposed to heavy artillery fire, further verifying Samuel's claim for a veteran disability pension. A letter from Samuel's doctor from May 22, 1889, indicates that Samuel "became totally deaf in Rt Ear 6 years ago. Has lost 2/3 hearing of left ear + at times been totally deaf in that ear. Expects soon to be totally deaf in both since left ear all the times growing worse". It further claims that Samuel "Has attacks every summer - Sometimes 2 or 3 times - accompanied by pain, swelling + Erysepilation inflammation of both eyes". "Erysepilaion" referring to erysipelas, the red inflammation of the outermost skin due to bacterial infection. Samuel was hence paid a "12/18 rating for the disability caused by deafness both ears, 4/18 for that caused by disease of eyes" according to his 1889 medical examination. Samuel's pension originally allowed only $20 per month (approximately $630 today), which would allow him to cover personal care costs, as well as any medical examinations or surgeries he may have needed in the future. Medical costs, however, only increased as both Samuel and his wife Lucy got older. By 1911, Samuel tried to apply for $40 in total benefits per month (approximately $1,250 today), as his medical bills for both him and his wife were starting to add up, Lucy being bedridden with cancer and Samuel no longer being able to work as an architect and carpenter due to his vision and hearing disabilities from the war. Samuel was forced to mortgage his property to pay for their hospital bills. Samuel was denied a $20 increase in his pension in 1911 initially. Later, Samuel received an increase of $10 to his pension, giving him $30 total each month. Due to his vision and hearing disability Samuel was forced to resign from his job as an architect and carpenter in Albion and was relegated to living at home with his wife in their final years.

By 1911 Samuel's wife Lucy Ann Saunders had cancer for close to three years and was in a more terrible condition than years previous. According to Lucy's obituary in Seventh Day Baptist journal *The Sabbath Recorder* she "bore her pain with such grace and cheerful fortitude that it awakened the amazement of all who knew her".[1] Samuel and Lucy moved to the capital of Madison just west of Albion in 1911 according to Samuel's obituary. After the death of Samuel's wife Lucy in 1912 Samuel lived alone as a widower in his apartment located at the Bellevue Court apartments in south Madison. Samuel ultimately died from heart failure on the steps of his apartment in Bellevue Court in Madison on September 28, 1914.[2] Samuel's daughter Mae Burdick-Bolser had found

---

[1] *The Sabbath Recorder*, Volume 72, No 8, Feb. 19, 1912, 252.

[2] *Wisconsin Tobacco Reporter*, October 2nd, 1914, Image 5. *Chronicling America, Library of Congress*.

Samuel Burdick Passes Away at Madison.

Samuel Burdick, a civil war veteran, was found dead on the stairs leading to his home in Bellevue Court, South Madison, Monday. Death was evidently due to heart failure. Mr. Burdick was 81 years old and had lived in Madison for the last three years.

A daughter, Mrs. M. J. Bolser, found the body in sitting posture. It is believed death came shortly before 11 o'clock.

The body was shipped to Albion, at which place burial took place Wednesday afternoon. The deceased is survived by a daughter in Madison and two brothers, A. C. Burdick of Albion and N. M. Burdick of Providence, R. I.

(l) Samuel's obituary as seen in the *Wisconsin Tobacco Reporter* from October 2, 1914. *Courtesy of the Library of Congres*; (r) Samuel Burdick Jr. and his wife Lucy Ann Saunders's headstone at Evergreen Cemetery in Albion. Buried across from Samuel is his brother Albert C. Burdick (1840-1919) with his two wives Elnora "Nora" P. Coon (1848-1883) and Dora Emelia Webster (1863-1945). *Courtesy of Roy C. Whitford.*

his body while coming to visit him in Madison. Samuel's body was transported back to Albion where he was buried in Evergreen Cemetery beside his wife and surrounded by friends, his younger brother, and extended family.

Samuel Burdick's experiences of the Civil War is just one of hundreds of thousands of men that served during an era of turmoil in United States history. The American Civil War is still to this day the United States' bloodiest war. Although the war only lasted roughly four years from 1861 to 1865, it cost more than roughly 800,000 lives of both northern and southern soldiers. This number of American casualties has still not been surpassed since the American Civil War. This annotated diary was created in order to share these experiences, as well as provide context to the rough life of the average Wisconsin soldier during the American Civil War directly from a soldier's perspective.

The 23rd Wisconsin Infantry Monument at Vicksburg.
Courtesy of the Historical Marker Database.

Epilogue

Notes on the 23rd Wisconsin
Volunteer Infantry Regiment

The 23rd Wisconsin Infantry Regiment had an original strength of 994 soldiers enrolled in its ranks when it was first mustered into federal service on August 30, 1862 at Camp Randall in Madison, Wisconsin.[1] According to E.B. Quiner's *Military History of Wisconsin*, the regiment gained one recruit in 1863, one hundred and eighteen recruits in 1864, and four recruits in 1865, totaling the regiment to 1,117 soldiers at the height of its strength.[2] Hosea W. Rood's *Wisconsin at Vicksburg: Report of the Wisconsin-Vicksburg Monument Commission*, also states the following statistics; original strength of regiment: 994, number of recruits received 123, total strength: 1,117.[3] Of the 1,117 men who served in the 23rd Wisconsin between 294, 308, and 316 men died in total dependent on the source.[4] One source, Charles E. Estabrook's *Wisconsin Losses in the Civil War*, states that the 23rd Wisconsin lost 294 soldiers in total; twenty-two were killed in action, sixteen died of their

---

[1] Quiner, *The Military History of Wisconsin*, 719; Ronald Paul Larson, *Wisconsin and the Civil War*, (Charleston: The History Press, 2017), 23.

[2] Quiner, *The Military History of Wisconsin*, 719.

[3] Hosea W. Rood, *Wisconsin at Vicksburg: Report of the Wisconsin-Vicksburg Monument Commission Including the Story of the Campaign and Siege of Vicksburg in 1863: With Especial Reference to the Activities Therein of Wisconsin Troops* (Madison: Wisconsin-Vicksburg Monument Commission, 1914), 161.

[4] Quiner, *The Military History of Wisconsin*, 719.

wounds, two hundred and fifty died of disease, and six died from accidents.[5] Contesting this statistic is William F. Fox's *Regimental losses in the American Civil War, 1861-1865* which states that the 23rd Wisconsin's casualties were as follows; "killed or died of wounds"; one officer, forty enlisted, "died of disease, accidents, in prison, etc."; five officers, 262 enlisted men, with a total of 308 casualties altogether.[6] Lastly, Hosea W. Rood's *Wisconsin at Vicksburg*, also lists the casualties of the 23rd Wisconsin as; Killed and died of wounds; forty-one died of disease; 262, died of accidents; thirteen with a total loss of 316 men.[7] It cannot be said for certain, but the 23rd Wisconsin Volunteer Infantry Regiment based of these three secondary sources lost roughly three hundred men according to most sources.

The 23rd Wisconsin would go on to be commemorated in stone and steel through a variety of memorials and monuments in the latter half of the nineteenth century, primarily in its most famous engagement, the Siege of Vicksburg which is now the Vicksburg National Military Park. Four major monuments, statues, or plaques still stand in Vicksburg today that relate to or commemorate the 23rd Wisconsin Infantry Regiment, those being the William F. Vilas statue, the Wisconsin Monument, the 23rd Infantry Memorial, and the 23rd Wisconsin Infantry Regiment Camp plaque.

The Vilas Memorial at Vicksburg. *Wisconsin Veterans Museum.*

---

[5] Charles E. Estabrook, *Wisconsin Losses in the Civil War: A list of names of Wisconsin soldiers killed in action, mortally wounded or dying from other causes in the Civil War. Arranged according to organization and also in a separate alphabetical list, 1915*, (Madison: Adjutant Generals Office, 1915), 121.

[6] William F. Fox, *Regimental losses in the American Civil War, 1861-1865 :A treatise on the extent and nature of the mortuary losses in the Union regiments, with full and exhaustive statistics compiled from the official records on file in the state military bureaus and at Washington*, (Albany: Albany Publishing Company, 1889), 513.

[7] Hosea W. Rood's *Wisconsin at Vicksburg*, 161.

> ### Vilas Statue for Vicksburg.
>
> Mrs. William F. Vilas, widow of the late Col. William F. Vilas, it is announced, has decided to donate a standing statue of Col. Vilas to the Vicksburg National Park Commission. Col. Vilas took part in the memorable siege at Vicksburg. Details regarding the statue are not yet announced.

"Vilas Statue forVicksburg" from the *Wood County Reporterm* May 5, 1910. *Courtesy of the Library of Congress.*

The Colonel Vilas statue was first commissioned in 1909 after the death of Vilas one year prior in 1908. The statue was made by Adolph A. Weinman of New York City, New York, and was constructed of bronze on a granite pedestal and base. The statue was dedicated on November 12, 1912, at Vicksburg to Mrs. Vilas and Mrs. Hanks, Vilas' daughter.[8] According to Hosea W. Rood, the 23rd Wisconsin monument is located on "Union Avenue a little North of the Balwin's Ferry Road" at Vicksburg National Military Park.[9] This location was chosen because "the regiment in the lines was a little north of the railroad leading to Jackson near the present Hebrew cemetery".[10]

In 1901 a board of commissioners was created by the Governor of Wisconsin Robert M. La Follette in order to erect and establish a dedicated war memorial to all Wisconsin regiments present at the Siege of Vicksburg. By 1903 the Wisconsin Legislature voted to appropriate $30,000 to erect a dedicated memorial. Ultimately the commission decided to erect a memorial made from South Carolina Winnsboro Granite and contracted the memorial through the Harrison Granite Company of New York. The architect of the memorial was William Liance Cottrell (1868-1964) of Westerly, Rhode Island and the sculptor was Julius C. Loester (1860-1923) of New York.[11] Interestingly enough, William L. Cottrell was a distant cousin of the previously mentioned Sergeant Zerah Cottrell in Samuel's diary. The monument was dedicated on May 22, 1911, on the 48th anniversary of the storming of

---

[8] Rood, *Wisconsin at Vicksburg*, 28.

[9] Rood, *Wisconsin at Vicksburg*, 120-121.

[10] Ibid.

[11] Ibid., 10-12.

the Vicksburg defenses to a large crowd.[12] Around to the balustrade of the monument are fifteen bronze tablets, each of which contains the names of all 9,059 Wisconsin soldiers who took part in the Vicksburg Campaign and the Siege of Vicksburg. The list consists of the following units from Wisconsin:

8th Wisconsin Infantry Regiment: 593 men.

11th Wisconsin Infantry Regiment: 625 men.

12th Wisconsin Infantry Regiment: 844 men.

14th Wisconsin Infantry Regiment: 437 men.

16th Wisconsin Infantry Regiment: 415 men.

17th Wisconsin Infantry Regiment: 529 men.

18th Wisconsin Infantry Regiment: 360 men.

20th Wisconsin Infantry Regiment: 605 men.

23rd Wisconsin Infantry Regiment: 858 men.

25th Wisconsin Infantry Regiment: 830 men.

27th Wisconsin Infantry Regiment: 715 men.

29th Wisconsin Infantry Regiment: 738 men.

33rd Wisconsin Infantry Regiment: 702 men.

2nd Wisconsin Cavalry Regiment: 444 men.

1st Independent Battery Wisconsin Light Artillery: 139 men.

6th Independent Battery Wisconsin Light Artillery: 127 men.

12th Independent Battery Wisconsin Light Artillery: 114 men.[13]

---

[12] Ibid.

[13] Rood, *Wisconsin at Vicksburg*, 372-373, 501.

The 23rd Wisconsin Infantry Camp marker at Vicksburg.
Courtesy of the Historical Marker Database.

"Wisconsin Memorial on Vicksburg Battleground" from *The Baraboo News*,
May 25, 1911. *Courtesy of the Library of Congress.*

Bibliography

Beck, J. D., *The Blue Book of the State of Wisconsin*. Madison: Democrat Printing Company, 1907.

"Benjamin Hyde Edgerton: Wisconsin Pioneer." *The Wisconsin Magazine of History* 4, no. 3 (1921): 354–57. https://www.jstor.org/stable/4630316

Billings, John B. *Hardtack and Coffee: The Unwritten Story of Army Life*. Boston: George M. Smith & Co., 1887.

Carley, Kenneth. *Minnesota in the Civil War: An Illustrated History*. St. Paul: Minnesota Historical Society Press, 2000.

Carley, Kenneth. *The Dakota War of 1862: Minnesota's Other Civil War*. St. Paul: Minnesota Historical Society Press, 1976.

Chambers, Whiteclay John, Anderson, Fred, *The Oxford Companion to American Military History*. New York: Oxford University Press, 1999.

Champlin, John Denison. *Young Folk's History of the War for the Union*. New York: H. Holt &Company. 1881.

Cook, Joel. *The Siege of Richmond: A Narrative of the Military Operations of Major-General George B. McClellan During the Months of May and June, 1862*. Philadelphia: G. W. Childs, 1862.

Cox, Jacob D. *The March to the Sea; Franklin and Nashville*. New York: C. Scribner's Sons, 1898.

Current, Richard N. *The History of Wisconsin, Vol II: The Civil War Era.* Madison: Wisconsin Historical Society, 1976.

Davis, William C. *Rebels & Yankees: Battlefields of the Civil War.* London: Salamander Books Ltd., 1991.

Dyer, Frederick H. *A Compendium of the War of the Rebellion: Compiled and Arranged from Official Records of the Federal and Confederate Armies, Reports of the Adjutant Generals of the Several States, the Army Registers, and Other Reliable Documents and Sources.* Des Moines: Dyer Pub. Co., 1908.

Estabrook, Charles E. *Annual Reports of the Adjutant General of the State of Wisconsin for Years 1860-1865.* Madison: Democrat Printing Co., 1860.

Fisk, Wilbur. *Hard Marching Every Day: The Civil War Letters of Private Wilbur Fisk, 1861-1865.* Lawrence: University of Kansas Press, 1992.

"Gen. Joshua Guppy Dead." *Watertown Republican*, December 13, 1893. https://chroniclingamerica.loc.gov/lccn/sn85033295/1893-12-13/ed-1/seq-6/

Hayes, Martin A. *A History of the Second Regiment, New Hampshire Volunteer Infantry, in the War of the Rebellion.* Lakeport: Unknown, 1896.

Jones, M. E. *The Squirrel Hunters of Ohio: or Glimpses of Pioneer Life.* Cincinnati: The R. Clarke Co., 1898.

Kent, Frank L., *Wisconsin in the Civil War: The Home Front and the Battle Front, 1861-1865,* Madison: The State Historical Society of Wisconsin, 1997.

Lerwell, Leonard L. *The Personnel Replacement System in the United States Army.* Washington, D. C.: Department of the Army, 1954.

Ligowsky, A. *Map of Dane County, Wisconsin.* Madison: Menges and Ligowsky, 1861.

Love, William De Loss. *Wisconsin in the War of the Rebellion; a History of All Regiments and Batteries the State Has Sent to the Field, and Deeds of Her Citizens, Governors and Other Military Officers, and State and National Legislators to Suppress the Rebellion.* Chicago: Church and Goodman, 1866.

McCarthy, Dorothy G. "The Contributions of Joshua Guppey," *Portage Daily Register*, January 22, 1972. https://www.newspapers.com/clip/78850376/joshua-james-guppey-1820-1893/

McPhearson, James M. *Battle Cry of Freedom: The Civil War Era*. New York: Ballantine Books, 1988.

Moe, Richard. *The Last Full Measure: The Life and Death of the First Minnesota Volunteers*. St. Paul: Minnesota Historical Society Press, 2001.

Monroe, Marie Jussen. "Biographical Sketch of Edmund Jussen." *The Wisconsin Magazine of History* 12, no. 2 (1928): 148-149, 159, 165, 167, 171, 175, 231. https://www.jstor.org/stable/4630753

Murdock, Eugene Converse. *Patriotism Limited, 1862–1865: the Civil War Draft and the Bounty System*. Kent: Ohio State University Press, 1967.

"Obituary – Burdick," The Wisconsin Tobacco Reporter, February 09, 1912, https://chroniclingamerica.loc.gov/lccn/sn86086586/1912-02-09/ed-1/seq-8/

Patterson, Gerard A. *Rebels from West Point*. Mechanicsburg: Stackpole Books, 2002.

Puck, Susan T. *Sacrifice at Vicksburg: Letters from the Front*. Shippensburg: Bird Street Press, 1997.

Quiner, E. B. *The Military History of Wisconsin: a Record of the Civil and Military Patriotism of the State, in the War for the Union, with a History of the Campaigns in Which Wisconsin Soldiers Have Been Conspicuous, Regimental Histories, Sketches of Distinguished Officers, the Roll of the Illustrious Dead, Movements of the Legislature and State Officers, etc.* Chicago: Clarke & Co., 1868.

Roster and record of Iowa soldiers in the War of the Rebellion, Together with Historical *Sketches of Volunteer Organizations, 1861-1866, Vol. III, 17th-31st Regiments-Infantry*. Des Moines: Emory H. English, 1908.

Rusk, Jeremiah M., and Chandler P. Chapman. *Roster of Wisconsin Volunteers, War of the Rebellion, 1861-1865. Volume I.* Madison: Democrat Printing Company, 1886.

Rusk, Jeremiah M., and Chandler P. Chapman. *Roster of Wisconsin Volunteers, War of the Rebellion, 1861-1865. Volume II*. Madison: Democrat Printing Company, 1886.

Sherman, William T. *Memoirs of General William T. Sherman*. New York: D. A. Appleton and Company, 1886.

Sobel, Robert; Raimo, John W. *Biographical Directory of the Governors of the United States, 1789–1978, Volume IV*. Westport: Meckler Books, 1978.

Strait, N. A. *Roster of All Regimental Surgeons and Assistant Surgeons in the Late War, With their Service, and Last-Known Post-Office Address*. Washington: N. A. Strait, 1882.

Thomas, Lorenzo. *General Orders No. 15*. Washington, D. C, War Department, 1861.

Tucker, Phillip Thomas. "Reconnaissance in Tensas Parish, April 1863: Missouri Confederates in Louisiana." *Louisiana History: The Journal of the Louisiana Historical Association* 31, no. 2 (1990): 193-197, 205-206. https:// www.jstor.org/stable/4232791

Tucker, Phillip Thomas. *The Forgotten "Stonewall of the West": Major General John Stevens Bowen*. Macon: Mercer University Press, 1997.

Index

DANIEL SCHARFENBERG is a military historian, museum registrar, and family genealogist hailing from Minnesota. He received an M.A. in Public History at the University of Nebraska-Kearney and a B.A. in History from Bemidji State University.

His interest in history stems from his family and their lived experiences through significant events in European and United States history. His family traces its origins to Western Europe and has been on the American Continent since the arrival of the Mayflower in 1620, with his 14th Great-Grandfather Elder William Brewster IV of Derbyshire, England, as well as his 9th Great-Grandfather Robert Burdick of Devon, England. Robert Burdick was later an instrumental figure in the settlement of Westerly, Rhode Island. Daniel's family came to the Midwest after fighting in the American Civil War seeking new beginnings on the frontiers of Illinois, Wisconsin, and Minnesota.

His specialties in the field of history consist of American and European military history, specifically in 19th and 20th-century warfare. Other topics of interest and study that he has worked on include the study of American material history, weapons, textiles, international relations, religion, American politics, and geography. During his downtime he spends time with his family, reenacts, travels, writes, and reads historical nonfiction. His next planned project is to publish a historical nonfiction narrative about Hatch's Battalion of Minnesota Cavalry during the Dakota War, as well as publish his Great-Granduncle Lawrence Theodore Burdick's diary and letters from the Second World War.